FIGHT GLOBAL WARMING

29 Things You Can Do

Sarah L. Clark

ENVIRONMENTAL DEFENSE FUND
New York

CONSUMER REPORTS BOOKS
A Division of Consumers Union
Yonkers, New York

Copyright © 1991 by Environmental Defense Fund
Published by Consumers Union of United States, Inc.,
Yonkers, New York 10703
All rights reserved, including the right of reproduction in whole or in part in any form.

Library of Congress Cataloging-in-Publication Data
Clark, Sarah.
    Fight global warming : 29 things you can do / Sarah Clark.
      p. cm.
    Includes bibliographical references and index.
    ISBN 0-89043-444-1 (Consumer Reports Books : pb)
    1. Global warming.   2. Greenhouse effect. Atmospheric.
3. Greenhouse gases.   4. Environmental protection.   I. Title.
QC981.8.G56C53   1991   91-22694
363.73'87—dc20   CIP
                                                                        Rev.

Design by Joy Taylor
First printing, September 1991
Manufactured in the United States of America
Printed on recycled paper
Permissions appear on page iv.

Fight Global Warming: 29 Things You Can Do is a Consumer Reports Book published by Consumers Union, the nonprofit organization that publishes *Consumer Reports,* the monthly magazine of test reports, product Ratings, and buying guidance. Established in 1936, Consumers Union is chartered under the Not-For-Profit Corporation Law of the State of New York.

    The purposes of Consumers Union, as stated in its charter, are to provide consumers with information and counsel on consumer goods and services, to give information on all matters relating to the expenditure of the family income, and to initiate and to cooperate with individual and group efforts seeking to create and maintain decent living standards.

    Consumers Union derives its income solely from the sale of *Consumer Reports* and other publications. In addition, expenses of occasional public service efforts may be met, in part, by nonrestrictive, noncommercial contributions, grants, and fees. Consumers Union accepts no advertising or product samples and is not beholden in any way to any commercial interest. Its Ratings and reports are solely for the use of the readers of its publication. Neither the Ratings nor the reports nor any Consumers Union publications, including this book, may be used in advertising or for any commercial purpose. Consumers Union will take all steps open to it to prevent such uses of its materials, its name, or the name of *Consumer Reports.*

Contents

Acknowledgments v
Introduction 1

1 | Global Warming: The Basics 3
The Greenhouse Century 4 • The Greenhouse Effect 6 • Models of Tomorrow 15 • Buying Time 18 • A Warmer World 20 • Atmospheric Wild Cards 21 • What We Must Do 27

2 | Slowing Global Warming 31
High-Leverage Options 32 • International Policies 36 • Goals for the United States 38 • Reducing Carbon Dioxide Emissions 39 • Reducing Other Greenhouse Gases 66 • Global Cooperation 70

3 | What You Can Do to Beat the Heat: 29 Steps 73
On the Road 76 • In the Home 78 • Tree Planting 93 • At Work 94 • Vote Green 95

Appendix
Publications and Information Services 97

Index 103

The New York–based Environmental Defense Fund is a member-supported environmental group whose scientists, economists, and attorneys develop practical solutions to environmental problems. Founded in 1967 in the successful battle to ban DDT, the Environmental Defense Fund has grown to become one of the nation's largest and most effective environmental organizations, with seven regional offices and more than 200,000 members. For membership information, please write: EDF Membership, 257 Park Avenue South, New York, NY 10010.

Figure 1 is illustrated by C. D. Keeling and T. P. Whorf. From Boden, T. A., P. Kanciruk, and M. P. Farrell, *Trends '90: A Compendium of Data on Global Change* (Oak Ridge, Tenn.: The Carbon Dioxide Information Analysis Center, Oak Ridge National Laboratory, 1990).

Figure 2 is illustrated by C. D. Keeling and G. Marland. From Boden, T. A., P. Kanciruk, and M. P. Farrell, *Trends '90: A Compendium of Data on Global Change* (Oak Ridge, Tenn.: The Carbon Dioxide Information Analysis Center, Oak Ridge National Laboratory, 1990).

Figures 3 and 4 are courtesy of the Intergovernmental Panel on Climate Change, World Meteorological Organization/United Nations Environment Programme. From Houghton, J. T., et al., eds., *Climate Change: The IPCC Scientific Assessment* (New York: Cambridge University Press, 1990). Reprinted with permission.

Figure 5 is reproduced from *Policy Options for Stabilizing Global Climate* (draft) (U.S. Environmental Protection Agency, 1991). Reprinted with permission.

The table on pages 75–76 is adapted from Udall, James R., "Turning Down the Heat," *Sierra* (July/August 1989). Reprinted with permission.

The table on pages 89–90 is adapted from Weiss, P., "Lighting the Way Towards More Efficient Lighting," *Home Energy* (January/February 1989). Reprinted with permission.

Acknowledgments

THE AUTHOR is indebted to the expertise and assistance of many individuals, without whom this book would not have been possible. Special thanks and gratitude go to Dr. Michael Oppenheimer, EDF Senior Scientist, for helping conceive this project and providing guidance throughout the writing and editing process. He patiently read numerous drafts and freely gave feedback and encouragement. Credit for the genesis of Part Three goes to free-lance writer William Goodman, who researched and wrote the earliest version.

Many other EDF staff provided source materials, comments, and answers to numerous questions. Their assistance and input were invaluable. They include Navroz Dubash, Dr. Diane Fisher, Dr. Rodney Fujita, Dr. Stuart Gaffin, Dr. Rebecca Goldburg, Scott Hajost, Dan Kirshner, Fred Krupp, Alice LeBlanc, Roger Pasquier, Dr. Stephanie Pfirman, Bruce Rich, John Ruston, Robert Yuhnke, former staff member Peter Miller, and board member Dr. Charles F. Wurster. Joel Plagenz deserves many thanks for seeing this project through the publishing process. Tracy Linder, Jim Ricketts, and Serge Riou provided crucial support. Several editors and reviewers at Consumer Reports Books were instrumental in the editing process.

Numerous persons, institutions, and agencies were generous with their time and resources. I am indebted to the Carbon Dioxide Information Analysis Center of Oak Ridge National Laboratory, the Environmental Action Foundation, Friends of the Earth International, NASA's Goddard Institute for Space Studies, ICF, Lawrence Berkeley Laboratory, and the Rocky Mountain Institute for their assistance.

This project of EDF's Global Atmosphere and Environmental Information Exchange programs was generously supported by the William Bingham Foundation, the George Gund Foundation, W. Alton Jones Foundation, Inc., the Joyce Foundation, Joyce Mertz-Gilmore Foundation, and Public Welfare Foundation, Inc.

Last, thanks are due to the loyal members of the Environmental Defense Fund, to whom this book is dedicated.

Introduction

IN THE SPACE of one summer during 1988, the greenhouse effect became a household phrase and the threat of global warming emerged from relative obscurity to the top of the world's environmental agenda. The front page and nightly news spread the word that human activities were altering the Earth's atmosphere, with potentially worldwide and irreversible consequences.

The *enhanced* greenhouse effect, by which emissions like carbon dioxide and other gases will eventually cause global warming, indeed raises serious questions. What are the exact causes? Where's the proof that global warming will occur? What are the likely effects? This book addresses these and other questions, so that readers can understand the science behind the public policy debate facing world leaders.

Understanding global warming yields more than an appreciation of scientific theory. It leads directly to the conclusion that we must start acting now. The "lag time" during which gases linger in the atmosphere, for example, means that today's greenhouse gas emissions will surround the Earth for decades. Only by considering the future consequences can we create and support the right environmental policies for today.

Apart from explaining the science and politics of global warming, however, this book addresses the most important question of all: What can I, as an individual, do about it? The answer is found in the 29 steps outlined to reduce the emission of pollutants that cause global warming. Without requiring major personal sacrifices, these steps range from using compact fluorescent light bulbs and energy-efficient appliances to voting for environmentally responsible elected officials.

Global warming is an international problem. Not only the United States, but all the world's nations must grapple with the task of cutting down emissions that cause global warming. One of the biggest benchmarks on the road to finding an international solution will be convincing developing countries to avoid our mistakes and go directly to new, efficient energy sources, such as solar and biomass. Such suggestions will fall on deaf ears if the United States doesn't make a major effort to reduce its thirst for, and wasteful use of, fossil fuels.

The United States government has been reluctant to take a leadership role in the international movement to slow global warming. Instead, it has proceeded on a business-as-usual course, postponing any commitment to concrete action. This book shows why we must start acting now. By following these 29 steps, we Americans can lead our elected officials toward policies that will guarantee a habitable Earth for future generations.

1 | Global Warming: The Basics

IT DIDN'T HAVE the impact of the drought and heat wave of 1988, when grain withered in midwestern fields. And it didn't have the horrific drama of 1989's Hurricane Hugo, slamming into the Carolinas with the fury of a hundred-year storm. Rather, 1990 managed to be the warmest year on record in an almost quiet way. The year did not feature a string of blistering July days, but a pleasant Indian summer in much of the nation. The winter was so mild that in the Northeast, flowering bushes set their buds in January. Nevertheless, a warm year should not have surprised anyone; the average temperature has been getting hotter over the last century. Perhaps this was just a harmless extension of that trend, perhaps it was just a natural fluctuation in the ever-unpredictable weather. Climate has its ups and downs, due to nothing more sinister than solar activity or more controllable than the earth's orbit around the sun. Perhaps this was just one of those ups.

But perhaps not. Dr. James Hansen of NASA's Goddard Institute for Space Studies has found that, based on global surface-temperature data, the seven warmest years since 1880 have all occurred between 1980 and 1990. Clearly the globe is warming, and with every record more and more scientists and policymakers have become increasingly suspicious that the cause isn't natural variability, but an enhanced greenhouse effect. For years scientists have been warning that the gases released into the atmosphere by power plants, automobiles, agriculture, and industry threaten to alter the world's climate radically, raising average temperatures enough to precipitate heat waves and droughts and raise the levels of the oceans until coastlines from Bangladesh to Malibu are inundated. Whether or not scientists will look back on the preceding decade and decide that it amounted to an unambiguous signal that the long-awaited enhanced greenhouse effect had arrived, one thing seems clear: from now on, we can look forward to a fast-changing world that will bring aberrations not nearly as benign as the weather of 1990. If the average world temperature rises by even the few degrees that scientific consensus predicts it will, the twenty-first century could see not merely pleasant Septembers and relief from heating bills, but changes so sudden and dramatic that they may threaten food supplies, coastlines, the survival of species—and the ability of humans to eke out a living on a planet that may seem to have turned against them.

The Greenhouse Century

By the middle of the next century, both land and sea are expected to warm. Regions now covered with sea ice will be open waters. High latitudes, such as northern Europe and Canada, will experience more warming than equatorial regions. People living in coastal regions will suffer more and more flooding. In some areas, coastal agriculture will become a hopeless battle against incoming tides. Storm surges and flooding during cyclones and hurricanes will accelerate beach erosion and sweep away houses and roads; coral reefs and mangroves may disappear under the tepid rising tides.

In some cities, people will sweat through scores of days with temperatures above 95 degrees. Supplies of drinking water will become unreliable where spring snow-melts that now fill reservoirs become memories. Smog may take longer to disperse over urban areas. Torrid summers will send electricity demand soaring. Insects once known only to the tropics will migrate north, turning outdoor activities into running battles between arthropod and human; the new arrivals may have few natural predators in these latitudes.

Good agricultural soils and climates may no longer match. The perfect growing conditions for the grains that are the basis of much of Western agriculture, and feed a good part of the world, will migrate from the American corn-and-wheat belt into parts of Canada, where soil conditions may limit the size of harvests. For those farmers who manage to hold on in Kansas and Iowa, droughts and heat waves may be more frequent. As a result, food prices could soar; in some parts of the world, famines that once made headlines could become the rule rather than the exception. Biological diversity will continue to plummet. There will be, quite simply, less life on Earth.

And the worst of it? The knowledge that the changes that appear by 2050 or so will be but a harbinger of what is still to come. The gaseous pollutants that warm the world act slowly, over decades. Most of those that we have spewed into the atmosphere will not have exerted their effects for several decades. But they will. We cannot call them back.

Just like Ebenezer Scrooge on Christmas Eve, society has been given a glimpse of the future. Is this world that climatologists forecast the world as it *will* be, or the world as it *may* be? Is there any escape from the greenhouse future, or can we take steps now to mitigate, if not to prevent entirely, the consequences of a centuries-long habit of using the atmosphere as our sewer?

It is the premise of this book that there is still time. A certain amount of climate change is all but inevitable because of our past abuse of the atmosphere. But if the countries of the world heed the warning signals, much of the damage can be forestalled and, in time, even reversed.

6 The Greenhouse Effect

It won't be easy. There are years of accumulated damage to undo—damage that began with the Industrial Revolution of the late 1700s and early 1800s. Before then, the gases that make up the atmosphere had not varied by much for thousands of years (something scientists know by examining the tiny bubbles of ancient air trapped in glacial ice). The emission of gases from such natural sources as decomposing vegetation, which gives off carbon dioxide, was balanced by absorption of the gases by growing plants and other natural sinks, such as the oceans. Thus the atmospheric concentration of these so-called trace gases remained roughly constant.

Wood burning marked the beginning of human impact on the atmosphere. But then, in order to power its new steam engines and factories, industrializing society began burning coal with a vengeance. That added to the atmosphere large amounts of heat-trapping gases, primarily carbon dioxide. As a result, for the first time in the planet's history, one of its denizens—human beings—had the power and the hubris to change Earth's very climate. The growth in the atmosphere's concentration of greenhouse gases will lead, slowly but inexorably, to an increase in Earth's average temperature.

This warming trend is caused by the enhanced greenhouse effect. To understand how this works, imagine yourself walking into a garden greenhouse on a nippy November morning. The air inside is warmer than it is outside. The reason is that glass permits sunlight to pass through to warm the interior, then traps the heat in the sunshine-warmed air, deflecting it back inside. The greenhouse effect refers to an analogous process in which a variety of natural gases in Earth's atmosphere act like the glass in a greenhouse, warming the planet. The following table identifies the primary greenhouse gases and their sources, both man-made and natural.

Greenhouse Gases and Their Sources

GHG	Man-made Sources	Natural Sources
Carbon dioxide	Fossil-fuel combustion Burning of forests to clear land Cement manufacturing	Decomposition of organic material, such as dead trees
Chlorofluorocarbons	Refrigerants for air-conditioning and refrigeration Plastic foam production Electronic cleaning solvents Hospital sterilants	None
Methane	Rice cultivation Incomplete fossil-fuel combustion Cattle flatulence Burning of biomass, such as forests and cropland Coal mining Natural gas pipeline leaks Solid waste landfills Increased number of termites	Wetlands Ruminant animal flatulence Forest fires Termites Seepage from the earth
Nitrous oxide	Ammonium fertilizers Coal combustion Burning of biomass	Microbial activity in soil
Tropospheric ozone	Photochemical reactions b/w hydrocarbons and nitrogen oxides emitted by fossil-fuel combustion	Photochemical reactions w/ natural hydrocarbons and natural oxides

The atmosphere is largely transparent to incoming, short-wavelength sunlight. This radiation passes through the atmosphere to the planet's surface, where much of the energy it contains is absorbed by the ground and by water. Like any warm objects, these radiate the heat back toward space in the form of longer-wavelength "infrared" radiation. But much of the heat never escapes the atmosphere: some gases, primarily water vapor and carbon dioxide, trap infrared radiation and deflect it back toward the planet's surface. Only about one-third of the solar energy hitting our planet bounces back into space directly.

Without this greenhouse effect, Earth would be about 60 degrees Fahrenheit colder than it is and mostly covered by ice. Thus, it is actually the greenhouse effect that makes life on Earth possible.

The greenhouse effect is rapidly becoming a greenhouse problem because human activities are adding greenhouse gases to the atmosphere at a more rapid rate than natural sinks are able to absorb them. Some of these gases—carbon dioxide, methane (swamp gas), and nitrous oxide—already occur in nature. Other greenhouse gases are synthetic, primarily the chlorofluorocarbons (or CFCs, also known by one trade name, Freon) better known for eating a hole in the stratospheric ozone layer that screens out ultraviolet radiation from the sun.

Carbon Dioxide

Carbon dioxide (CO_2), a colorless, odorless gas, is the principal greenhouse gas. It is an unlikely villain, making up a mere .035 percent of the air we breathe. Yet that compares to .028 percent of the air at the dawn of the Industrial Revolution. That tiny percentage, translated into weight, tips the scales at about 750 billion tons of carbon dioxide. Today various human activities are spewing out so much carbon dioxide that the same amount that entered the atmosphere in the 100 years between 1850 and 1950 is being emitted every 12 or so years. It is difficult to recognize the erratic squiggle on scientific graphs tracing the steady rise in carbon dioxide over the last 200 years for the threat that it is. But the meaning is clear: carbon dioxide is thought to be responsible for more than half (55 percent) the greenhouse problem.

It is also the greenhouse gas that we know the most about, thanks in large part to a pioneering experiment on top of a Hawaiian volcano.

In 1958 Charles Keeling of the Scripps Institution of Oceanography began monitoring the atmospheric carbon dioxide concentration at the Mauna Loa Observatory in Hawaii. In that year, 315 ppmv (parts per million, by volume) of carbon dioxide were present. The 1990 concentration was 353 ppmv. This represents an increase of 12 percent over the past 32 years, and is graphically illustrated in Figure 1.

A source of data about older carbon dioxide levels comes from air bubbles in glacier ice. Snowflakes trap air as they accumulate, air that remains isolated as bubbles once the snow turns to ice. Measurements of this "fossil air" show that the atmospheric concentration of carbon dioxide at around 1750 was 280 ppmv. The increase of 25 percent since preindustrial times—and of more than 10 percent in just the last 30 years—is a consequence of two factors: fossil-fuel combustion and deforestation.

How do scientists know this? Direct evidence that fossil-fuel combustion is largely responsible for the increased atmospheric carbon dioxide concentration comes from comparing the graphs of known human usage of fossil fuels (see Figure 2) to the Mauna Loa graph of atmospheric carbon dioxide concentration.

The close correspondence between the trends of these two graphs is very clear evidence that our energy usage is changing the composition of the atmosphere. This same comparison,

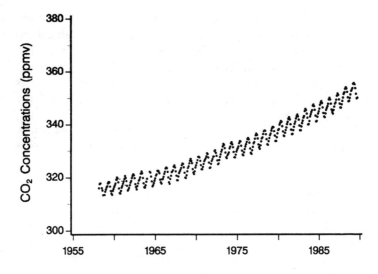

FIGURE 1. *Atmospheric carbon dioxide concentrations measured at the Mauna Loa Observatory (Hawaii, U.S.A.) between 1955–1989*

however, also reveals another interesting fact: less than half of the carbon dioxide released by combustion and deforestation in the tropics is actually showing up in the Mauna Loa measurements. Where is the rest of it going?

The combination of a rising concentration of gaseous carbon dioxide in ocean surface water, the growth of northern forests, and the standing concentration of carbon dioxide in the atmosphere (the Mauna Loa measurements) probably accounts for the carbon dioxide emitted by fossil-fuel combustion. Whether the oceans act as a net source or a sink of carbon dioxide for the atmosphere depends on the gas-exchange process, which in turn depends upon the varying ocean surface water concentration of gaseous carbon dioxide relative to how much is in the air above the oceans. Calculations have shown that, given the current state of the surface waters of the oceans, they

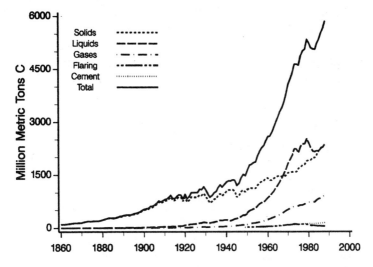

FIGURE 2. *Global carbon dioxide emissions from fossil fuel burning, cement manufacture, and gas flaring, between 1860–1988*

indeed act as a *net* sink for atmospheric carbon dioxide with a magnitude that can account for roughly one-third of the carbon dioxide released by fossil-fuel combustion and deforestation in the tropics. About half stays in the atmosphere, and the balance (about one-sixth of the total) presumably is sequestered in trees in the Northern Hemisphere.

The carbon budget. The so-called carbon budget—an accounting of the sources and sinks of carbon on Earth—is subject to considerable scientific uncertainty. But our best estimate is that we are now injecting close to 6 billion tons of carbon into the atmosphere every year through the burning of such fossil fuels as petroleum, coal, and natural gas. Clearing tropical and other forests adds another 1 to 2 billion tons or so of carbon per year. Of this approximately 7 billion tons annually, slightly over 3 billion tons remain aloft. The rest is absorbed into any of the planet's carbon sinks. This 3 billion tons of carbon with no place to go is the crux of the greenhouse problem.

According to analysis of the Vostok (Antarctic) ice core, atmospheric levels of carbon dioxide and world temperature have been inextricably linked over thousands of millennia. The bottom of the core was formed from snow that fell 160,000 years ago, and chemical measurements indicate how much carbon dioxide was present in the atmosphere at particular times as well as what global temperatures were like. (These past fluctuations in carbon dioxide were triggered by natural variations in Earth's orbit, which change the seasonal pattern of sunlight hitting the planet's surface and thus the rate of plant growth, the temperature of the oceans, and consequently the amount of atmospheric carbon dioxide.) The carbon dioxide increase inferred from the Vostok ice core altered the atmosphere enough to account for a good deal of the 5-degree-Celsius warming that occurred as the last ice age ended. Although some skeptics question whether changes in carbon dioxide levels cause temperature to vary, or whether changes in temperature (due to some other factor) affect carbon dioxide concentrations, the historical shifts in climate provide a natural experiment that supports the global warming theory.

Methane

Although increasing carbon dioxide levels are the greatest cause of concern, atmospheric concentrations of some other greenhouse gases in the atmosphere are rising even more quickly. Levels of methane in the atmosphere have doubled since preindustrial times, from .8 parts per million to 1.72. Molecule for molecule, these other greenhouse gases cause even more warming than carbon dioxide does. One molecule of methane (the main constituent of natural gas) is about 21 times as effective as a molecule of carbon dioxide in retaining heat in the atmosphere because of its greater capacity to absorb infrared radiation. So even though it is present in the air in far smaller quantities than carbon dioxide, methane is responsible for about 15 percent of the greenhouse problem.

Researchers speculate that atmospheric methane is increasing because its sources have been steadily growing and its sink may be decreasing. Hydroxyl radicals (generated by a complex set of reactions) react chemically with methane in the atmosphere, acting as the major natural methane sink. Carbon monoxide and methane pollution may have depleted hydroxyl concentrations in the atmosphere, causing this sink to have a diminished capacity to remove methane.

Natural sources of methane include organic-rich sediments in the shallow waters of wetlands, tundra, and lakes; the digestive tracts of ruminant animals; seepage from the earth; forest fires; and termites. In addition, a large amount of methane is buried under the ocean in frozen deposits called methane clathrates, which are a potential source of methane if the oceans warm enough to melt these deposits. Sources associated with human activity include rice cultivation; natural gas systems; cattle, sheep, and goat ranching; petroleum extraction; biomass burning; coal mining; and solid waste landfills.

Chlorofluorocarbons

Chlorofluorocarbons (CFCs) are a family of synthetic chemicals first invented in the 1930s. Nontoxic and nonflammable, these "miracle" chemicals have a wide variety of uses: coolants in refrigeration and air-conditioning units, propellants in aerosols, blowing agents for foam products, cleaning solvents, and hospital sterilants. Depending on the use, CFCs are released to the atmosphere immediately, such as when computer components are degreased, or slowly over a product's lifetime, such as from the "closed" cells in residential insulation urethane foam.

In the early 1970s scientists suggested that CFCs destroy the ozone gas that encircles the earth in the upper layer of the atmosphere called the stratosphere. Ozone in the stratosphere filters out the Sun's harmful ultraviolet rays. Now we know that CFCs do in fact destroy ozone and are the cause of both the 4 to 5 percent global loss of ozone and the large seasonal depletion of the ozone layer, centered over Antarctica, known as the ozone hole. The ozone hole proved that waiting too long to take action can have catastrophic results.

Aside from being ozone depleters, CFCs are potent absorbers of heat radiated from Earth—20,000 times more effective, molecule for molecule, than carbon dioxide. Two of the most important ozone-depleting CFCs, CFC-11 and CFC-12 (the names come from the arrangement of the atoms in the molecule), persist in the atmosphere for 65 to 110 years and currently account for 17 percent of the greenhouse problem. Other CFCs account for 7 percent, making a grand total contribution of 24 percent from CFCs.

Because humans are the only source of CFCs, it's easy to account for their buildup in the atmosphere and for their absence from any atmospheric records more than 50 years old. They currently grow at a rate of 4 percent a year.

Nitrous Oxide

Nitrous oxide (commonly known as laughing gas) accounts for 6 percent of the greenhouse problem. The atmospheric nitrous oxide content grew 3.5 percent between 1975 and 1985. Although this increase is less rapid than for many other greenhouse gases, nitrous oxide lasts longer in the atmosphere—roughly 150 years—than any of these gases besides carbon dioxide. Microbial activity in soils, which breaks down nitrogen compounds under natural conditions, is the principal source of nitrous oxide. The gas is also released when nitrogenous fertilizers are applied to soil and during biomass burning, causing the atmospheric concentration of nitrous oxide to rise. However, nitrous oxide's growth rate cannot be fully accounted for by these natural and man-made sources. Either one or more sources are missing, or the contributions of some sources are underestimated. Nevertheless, human activities are believed to be the cause of the increasing atmospheric concentrations of nitrous oxide.

Ozone

Ozone in the stratosphere shields Earth from ultraviolet radiation. But ozone in the lower atmosphere (troposphere) is a pollutant, irritating people's eyes and lungs, harming plants, and damaging materials. This tropospheric ozone is a principal component of smog. It's the end product of a complex series of sunlight-catalyzed reactions involving hydrocarbons and nitrogen oxides (produced mainly by automobiles).

Tropospheric ozone is twice a villain, as it is also a greenhouse gas. It's responsible for a small portion of greenhouse warming (scientists have not yet calculated its contribution precisely). If you're thinking, if only we could move ozone from ground level to the stratosphere to alleviate both global warming and the ozone hole, it just isn't economically or technically feasible.

The human contribution to tropospheric ozone has been difficult to assess in part because it is affected by depletion of ozone within the stratosphere. The tropospheric ozone concentration is increasing at about 1 percent a year in the Northern Hemisphere. In the Southern Hemisphere, where man-made sources of hydrocarbons and nitrogen oxides are much fewer, changes in atmospheric ozone concentrations are not appreciable.

Models of Tomorrow

Computer models called General Circulation Models predict a number of global consequences as a result of the buildup of these greenhouse gases. To construct a GCM, climatologists calculate averages for such things as temperature, humidity, cloudiness, and altitude (mountains and valleys) within boxes of a grid that represents Earth. The boxes are about 300 miles on a side, so the predictions of the computer models are rough and are made on such a large scale that they cannot be applied to any region smaller than a continent. (Japan and New Zealand are too small even to appear on them.) The scientists then insert into their mathematical equations values for the concentrations of greenhouse gases and let the supercomputer run, sometimes for months. It takes the computer that long to crank through the myriad calculations needed to predict the effect of the greenhouse gases on climate.

There are currently five different models (four written by scientists in the United States and one by climatologists in Great Britain), but none is able to predict how a particular country, much less a specific state or city, will fare in a greenhouse future. Nevertheless they all agree that Earth is on the verge of a warming of unprecedented magnitude.

The computer simulations of Earth's climate system indicate that a doubling of carbon dioxide from preindustrial levels—or an increase in all the greenhouse gases equivalent in climatic effect to a doubling of carbon dioxide—will lead to an increase in average global temperature of 1.5 degrees to 4.5 degrees Celsius compared to the mean temperature 100 years ago. Based on current trends of greenhouse gas emissions, Earth will continue to warm, by about .3 degrees Celsius (.54 degrees Fahrenheit) in each decade of the next century. According to the Intergovernmental Panel on Climate Change convened under the auspices of the United Nations Environment Program and the World Meteorological Organization, and reporting in 1990, the planet's mean temperature will rise above today's average temperature by about 1 degree Celsius (1.8 degrees Fahrenheit) by 2025, and more than 3 degrees Celsius (5.4 degrees Fahrenheit) before the year 2100.

There is a considerable range in the predicted warming effect because different climate simulations use different assumptions about such factors as "feedbacks" and the time it takes oceans to absorb excess heat. To get a feel for the range, consider Figure 3. It shows temperature curves developed by the United Nations panel, based on three different climate responses. It assumes a business-as-usual scenario; that is, society takes no steps to reduce the emissions of greenhouse gases. Given that assumption, estimates of the rise in average global temperature range between 2 and 7 degrees Celsius by the year 2100, depending on how climate responds to a doubling of carbon dioxide in the atmosphere. The most realistic estimate is a rise of 4 degrees Celsius over the preindustrial average.

Predictions vs. Reality

How well have the GCMs conformed to reality? The 1980s were clearly the decade of the greenhouse. Not only did public and political attention come to focus on this threat during the summer heat wave and drought of 1988, but, as previously stated, the last eleven years saw seven of the warmest years on record. A single hot year, or even six hot years, is no proof that the greenhouse crisis is upon us; there is too much normal variation in climate to attribute any particular short-term abnormality to the greenhouse effect. But looking backward produces a more worrisome conclusion. Temperatures are higher than they were a century ago. Climatologists at Goddard and at the Climatic Research Institute at the University of East Anglia in England have independently examined records from thousands of weather stations around the world. After correcting for the heat island effect, in which cities tend to be warmer than rural areas, both groups have documented a warming of about .5 degree Celsius, or .9 degree Fahrenheit, since records began being kept in the 1880s.

GCMs "predict" that the world warmed up about 1 degree Celsius in the past century, that is, plugging in the levels of

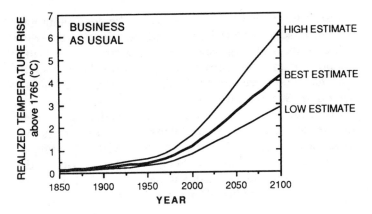

FIGURE 3. *Computer model predictions of global mean temperature under a "business-as-usual" scenario, with high-, low-, and best estimates of climate responses*

carbon dioxide in the air over the last 100 years produces a 1-degree rise in the mercury. That's about twice the actual change reported by the researchers at Goddard and East Anglia. A twofold discrepancy between theory and reality is not huge, considering the crudeness of the models, but it's still something to pay attention to. It could be that the state-of-the-art computer runs are too sensitive to greenhouse gases, that is, they calculate that these gases raise global temperatures more than they actually do. Or it could be that the models do not fully account for the capacity of the oceans to absorb heat, or for other mechanisms that counter greenhouse warming. Still, it seems reasonable to say that the predictions are probably reliable to within a factor of two.

Buying Time

The expected warming predicted by the models is extremely rapid when compared with most other climate fluctuations of the last 10,000 years. It is this—the *speed* of warming—that is as much the issue as the magnitude. Yes, world climate has changed before. Twenty thousand years ago, in the most frigid depths of the last ice age, much of the northern United States and Europe groaned under glaciers more than a mile thick, and sea levels were more than 300 feet lower than they are today. What is today the ocean fringe where toddlers splash on our seashores was then dry land that extended out miles at many sites.

On average, world climate changes very slowly, more slowly than the rate climatologists are now predicting will occur under the current onslaught of greenhouse gases. Because of these gases, the world will warm more quickly than it did on average during the last glacial retreat 10,000 to 15,000 years ago. If we could slow down the warming, we would buy time to adjust. Our botanists and agronomists would have time to develop new crop strains adapted to the precipitation and temperature regimes in a greenhouse grain belt. Our civil engineers would have time to construct new irrigation systems. Coastal cities could plan their shorelines to prevent any more construction on beaches; in the worst case, they could also lay plans for sea walls.

Natural systems, too, would have time to adjust. Forests have the ability to "migrate"—slowly shift north or south in response to environmental changes. But they do so only slowly, sending out a new seed slightly to the north of their current boundary, for instance, and waiting years for it to mature and begin establishing a new stand of trees. Coastal wetlands, which are crucial to the productivity of the world's fisheries, could also migrate inland if given decades instead of years. New predator-prey relationships could be established, slowing or preventing the invasion of tropical and insect-borne diseases expected to strike defenseless northerly climes as they heat up. The world has adjusted before to ice ages and warm interglacial spells. But now it no longer has the luxury of time.

If, on the other hand, climate change does occur 40 to 100 times faster than biological communities have experienced in the last million years (as seems possible under current projections), then few species would be able to disperse fast enough to keep up with the changing habitats. It will be as if their very homes were moving out from under them—their preferred rainfall pattern a few hundred miles north, the climate regime to which they became adapted over millennia a few score miles south. The loss of species and habitats may well be irreversible. Dr. Thomas Lovejoy, Assistant Secretary for External Affairs at the Smithsonian Institution, believes that if climate change were to occur at this pace, "we could forfeit the major part of our biological heritage."

The extent to which atmospheric concentrations of greenhouse gases, as opposed to such natural changes as volcanic eruptions and fluctuations in the Sun's output, have caused world temperatures to increase over the last 100 years remains a matter of scientific debate. While solar output variability could have contributed to the documented increase in global temperatures, it probably has not been the major factor. In fact, as greenhouse gas emissions increase with time, solar output variability will play a smaller and smaller role in contributing to global warming.

A Warmer World

A warmer world will see many other changes besides those in thermometer readings. Increased temperatures will bring a modest increase in global precipitation as well as changes in rainfall patterns. Most likely, dry areas will become drier and wet areas wetter. The reason is that wet regions could experience higher evaporation rates, causing more intense storms, rainfall, and flooding. Dry areas, on the other hand, have a limited amount of water to release. Local increases in temperature of as little as several degrees, for instance, could significantly decrease runoff in the arid Colorado Basin. Water quality will deteriorate as we discharge waste into decreased stream flows.

One way or another, global warming will also cause sea levels to rise. Just how much will depend on the roles played by three factors—thermal expansion of the oceans, melting of alpine glaciers, and melting of the Greenland ice sheet. These effects may be counteracted to some degree by growth of the Antarctic ice sheet due to increased precipitation. Global sea level has already risen over the past 100 years by four to eight inches. Because of uncertainties in estimating these factors, it is difficult to predict how much sea level will rise if emissions of greenhouse gases continue unabated. However, a conservative estimate is that by the year 2030, global mean sea level will be over half a foot higher than it is now, and by the end of the century a rise of two feet is projected. No matter what change in sea level occurs, beach loss will result (every foot rise causes about 100 feet of beach loss on the East Coast of the United States), salt water will intrude farther into estuaries, groundwater, and agricultural lands, wetlands will drown, and the inland reach of storm surges will increase.

Atmospheric Wild Cards

There are many uncertainties in these predictions. Some are inherent to any attempt to predict the future. Who can foresee whether a worldwide depression will so slow economic activity that emissions of industrial gases will plummet? Who can predict a technological breakthrough that will provide clean, abundant energy at no environmental cost or absorb atmospheric carbon dioxide?

Other uncertainties are specific to greenhouse science. First of all, there are missing data. Scientists have not gotten precise measurements of the concentrations in all parts of the atmosphere of water vapor, ozone, and aerosols, all of which can affect global climate. Aerosols—small solid or liquid droplets suspended in the air—reflect sunlight from the planet and thus exert a net cooling effect.

That raises the other class of uncertainties: atmospheric science is still not sophisticated enough to say with certainty how various feedbacks (linkages between one changing factor and other variables) will interact with the greenhouse gases to affect climate. Feedbacks have not been accurately incorporated into the computer models that predict temperature increases as a result of the buildup of greenhouse gases. Positive feedbacks would magnify the greenhouse effect; negative feedbacks would dampen the effect. One example of a positive feedback is increased evaporation (as a result of a warmer planet), which will make air more humid and trap more heat (since water vapor is a greenhouse gas). Another positive feedback would be a reduction in the amount of sunlight reflected back into space by snow and ice. As the ice melts in the warmer world of tomorrow, polar oceans as well as the planet generally will warm.

Clouds

The atmosphere is full of wild cards, and clouds are the biggest one. Theory holds that a warmer world would produce more water vapor (through evaporation of surface water) and hence more clouds, which, as anyone living in an overcast region can attest, generally cause cooling. But it is also possible that increased evaporation will cause more violent convection and thunderstorms over only a small region of the globe. In that case the rest of the world could be drier, and hence less cloudy and warmer. In addition, clouds also trap heat emitted from the surface below, somewhat as a blanket keeps in body heat, and that has a net heating effect. To further complicate things, low-level stratus clouds, as well as midaltitude storm clouds, tend to cool the planet more than they warm it through heat trapping; cirrus clouds high in the atmosphere tend to have a net heating effect.

Most computer models foresee a net warming effect from changes in cloudiness; in other words, changes in clouds would exacerbate the greenhouse effect. But that's an extremely tentative conclusion. Without accurate measurements of how cloud formation dovetails with world temperature changes, scientists will not know whether clouds magnify or mitigate greenhouse warming.

Marine Plankton

Another wild card comes in the form of marine plankton, the tiny free-floating microorganisms that inhabit the uppermost, warm layer of the seas. They incorporate dissolved carbon dioxide into their shells or body tissues. Fish eat the plankton, and eventually their waste products—containing the carbon—fall to the ocean floor. Most dissolves; some of it is eaten, to be returned to the food chain; and some is incorporated into sediment, where it is thus sequestered from the atmosphere. The unknowns in this process concern factors that might increase or decrease the amount of carbon dioxide taken up by the oceans. Winds, for instance, as well as dust and dirt blown into the seas, seem to increase the amount of carbon taken out of the air by plankton. Whether the warmer climate of tomorrow will be breezier, acting as a check on further carbon dioxide–forced warming, or calmer, encouraging the buildup of carbon dioxide in the air, is totally unknown. Another uncertain factor that could affect plankton growth is ocean temperature.

Unknown, too, is whether technological remedies (such as switching to solar energy) can spare the world's economies the difficulty of weaning themselves from the fossil fuels that spew out the lion's share of greenhouse gases. Planting trees is the simplest techno-fix and desirable on its own merits. Other schemes can be less benign.

One recent idea has been to make marine algae take up more carbon dioxide than they are currently. The hypothesis holds that those algae living in areas of the ocean that are deficient in iron, a nutrient, could consume more carbon dioxide if the waters were "fertilized" with iron. But recent calculations show that this "Geritol hypothesis" is overrated as a greenhouse savior. Calculations by Wallace Broecker of the Lamont-Doherty Observatory of Columbia University show that if world emissions of carbon dioxide continue to the point where atmospheric levels reach 600 parts per million, successful fertilization of the world's oceans with iron would at best reduce the increase in carbon dioxide levels by about 7 percent. Clearly, marine algae will not save us.

Marine algae aside, the role more generally played by the oceans in absorbing carbon dioxide is not well understood. Depending on how the oceans behave, global warming could be greater or less than computer models predict. As mentioned before, the oceans absorb one-third to one-half of the carbon dioxide emitted into the atmosphere. Depending on how they react to higher surface-water temperatures, that absorption rate could increase or decrease. That in turn would slow down or speed up the enhanced greenhouse effect.

What-Ifs

Finally, the computer models do not incorporate the impact of living, growing things on the greenhouse effect. Growing plants absorb carbon dioxide. That's why cutting down forests adds to the greenhouse problem—it removes a carbon dioxide sink. But neither climatologists nor botanists know if a more carbon-rich atmosphere will stimulate plants to grow and thus raise their intake of carbon dioxide; enough, perhaps, to put a brake on the runaway greenhouse. This may seem like a minor factor in calculating the greenhouse effect, but recent studies indicate that the living things in the Northern Hemisphere may absorb more human-made carbon dioxide than do the oceans, which are usually considered the world's principal carbon dioxide sink.

The bottom line is that there are considerable unknowns in greenhouse science. But as Jerry Mahlmann, director of the National Oceanic and Atmospheric Administration's Geophysical Fluid Dynamics Laboratory at Princeton University, puts it, "I can't think of any combination of uncertainties that could conspire to make the greenhouse problem go away."

Some people believe that a few degrees of warming is nothing to lose sleep over. Others rationalize that changes in temperature and precipitation will make some areas of the world winners even if others are losers—Siberia could become a breadbasket even if much of Bangladesh disappears under the tides. But exactly who will win and who will lose is uncertain. It must be remembered that the computer models are extremely crude, geographically, and the disruption in people's lives from just a few degrees of warming is likely to be enormous.

Thus the greenhouse threat forces society into the uncomfortable position of having to make decisions about impending catastrophe in the absence of scientific certainty. As Bertrand Russell wrote, "The most savage controversies are those about matters as to which there is no good evidence either way." The greenhouse quandary isn't as extreme as all that—there is quite compelling evidence, outlined earlier, that man-made emissions are altering world climate.

But there remains significant scientific and policy-making uncertainty. What if we did ban all chlorofluorocarbons and ordered utilities to stop burning the coal and oil that belch so much carbon dioxide, only to discover from the computer models of the twenty-first century that none of that was necessary? Maybe it would be more prudent to wait and see what happens—then build dikes when the waves start lapping at our coastal condominiums and put biotechnology to the task of breeding crops that can tolerate the new climate order. Unfortunately we cannot afford that luxury, and the poorer nations of the world can afford it even less.

There is another important point. The uncertainties about

global climate cut both ways. Yes, there may be cloud mechanisms that mitigate the effects of greenhouse gases on world temperatures; but it may also be that any changes in the atmosphere exert disproportionate effects on the ocean—the largest heat sink and a potent influence on both global and regional climate. There is some evidence from ancient climate records that such small atmospheric changes can radically alter ocean circulation patterns, possibly exacerbating any direct effects on the atmosphere. We need only recall the discovery of the ozone hole over Antarctica in 1985. Not even the staunchest chlorofluorocarbon foe—to say nothing of atmospheric scientists—foresaw this breach of the ozone layer. But it has now been attributed decisively to CFCs.

Twenty years ago climatologists thought that the greenhouse crisis would not hit with any force before the middle of the next century. They did not then recognize that a rapid rise in greenhouse gases other than carbon dioxide, especially chlorofluorocarbons and methane, was also occurring.

Earth is at present locked into a warming of 0.5 to 2.0 degrees Celsius from greenhouse gases already emitted, in addition to the rise of 0.5 degrees Celsius already recorded. If we delay action, this lag effect will build. More inevitable warming will occur—with uncertain and irreversible consequences—because once in the atmosphere, greenhouse gases are more or less there to stay. Atmospheric levels of several greenhouse gases, particularly carbon dioxide, will not revert to "normal" levels, even in response to large emission reductions, for hundreds of years. *Nevertheless, emission reductions can slow warming and eventually stabilize temperatures.*

The uncertainties and gaps in our knowledge do not call into question whether global warming will happen. They simply raise the questions of when and how much. Further research and analysis will enable us to make better predictions about the extent and timing of climate change and the consequent effects in particular regions. But more research won't alter the fundamental conclusion: *global warming due to greenhouse gases is expected to alter climate faster than any sustained natural change in the last 160,000 years, making the earth warmer during the twenty-first century than at any time since modern humans evolved.*

Can societies adjust to fast change without substantial dislocation? Can natural ecosystems survive? The more greenhouse gases we emit, the greater the likely disruption. Therefore, actions on the part of governments, industries, and individuals to minimize climate change must be taken as soon as possible.

Figure 4 shows four projections of temperature increases from the preindustrial global temperature average made by the United Nations panel. The projections are based on four different scenarios assuming progressively increased levels of controls: Business-as-usual (Scenario A) assumes a coal-intensive energy supply and only modest achievements in efficiency increases. Scenario B assumes a shift to lower carbon fuels and large efficiency increases. Scenario C assumes a shift toward fossil-fuel alternatives during the second half of the 21st century. Scenario D assumes a faster shift toward fossil-fuel alternatives. For each scenario, the projected temperature increase is quite different. In Scenario A, a 4-degree-Celsius rise is projected to occur by the year 2100; Scenario B projects a 3-degree rise; while Scenarios C and D project rises of 2.3 and 2 degrees Celsius respectively. This graph makes clear that policy choices can make a big difference.

What We Must Do

Experience shows that policies to alleviate environmental problems take years, even decades, to formulate, and an equivalent period to implement. It took nearly 10 years, for instance, to reach an international agreement on controlling CFC emissions to curb stratospheric ozone depletion. Unfortunately the Antarctic ozone hole and significant global depletion of the ozone

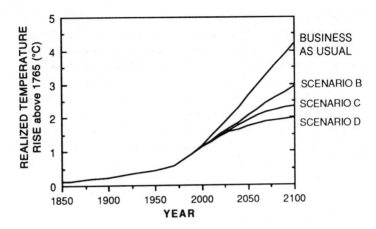

FIGURE 4. *Computer model simulations of global mean temperatures under four different policy scenarios*

layer may be the price we have paid for this delay. (Even if the ozone agreement had been reached in 1975, a few years after Dr. Sherwood Rowland and Dr. Ralph Molina of the University of California at Irvine discovered how CFCs could destroy ozone, but before the hole opened, the hole might still have appeared because of the slow migration of CFCs into the stratosphere over Antarctica.) If CFC production and use ceases by the year 2000, as the United States and other countries have agreed to, it will have been nearly 30 years from the discovery of stratospheric ozone depletion to the full implementation of a solution. Policy-makers consider the 1987 Montreal Protocol, which phases out worldwide CFC production, to have been achieved at breakneck speed.

Reaching a similar international agreement on greenhouse gases will be much more difficult, because doing so requires that the entire global energy system, and thus the economies of all nations, be revamped. But as we wait for politicians to realize the need for controls on greenhouse gases, we must not be misled into thinking that waiting—for more research, for technological fixes, for international consensus—is the most prudent course of action. In the case of the greenhouse, there is no "no action" option. Doing nothing does something: it allows greenhouse gases to continue to spew out of our cars, power plants, and factories—business as usual.

It is too late to prevent global warming. Some warming is already in the works. All we can do is slow it down. And for that, we must start now.

2 | Slowing Global Warming

THERE ARE all sorts of excuses not to address global climate change. The extent of warming over the past hundred years and the causes of it remain subject to (often rancorous) scientific debate; the climatic effects of adding greenhouse gases to the atmosphere are uncertain; the ultimate consequences of those climatic changes for life on Earth are unknown. This uncertainty provides some with a convenient excuse to forestall action to mitigate the greenhouse "until all the answers are in." This presumes that science can provide all the answers, which is an unrealistic expectation. While the countries of the world differ on how much to cut emissions of greenhouse gases, the United States has indicated a hesitancy to make new commitments because of uncertainties.

The second excuse for not taking action against the greenhouse is the hope of some countries that global warming will make them winners even if other nations are losers. Countries like Bangladesh and the Netherlands face great threats from rising sea levels, but regions where it is now too cold for much productive agriculture (Siberia, for instance, or northern Canada) can dream green dreams of fields of grain undulating over their greenhouse-warmed territory. This makes international consensus elusive.

Finally, many solutions to the greenhouse problem can be made to seem like prescriptions for economic disaster. Despite general public support for energy conservation and for rolling back environmental degradation, the cost of such measures can make prospective "green" taxpayers think again about their commitment to the environment. Some have issued glum warnings that controlling carbon dioxide and the other greenhouse gases will cripple productivity. Partly for these reasons, in the 1990 elections, Californians soundly defeated a voter initiative that would have, among other things, mandated cuts in the state's emissions of carbon dioxide and chlorofluorocarbons.

High-Leverage Options

The greenhouse crisis has one welcome characteristic: the policies available to slow global warming are "high-leverage" options. In other words, in addition to cutting emissions of greenhouse gases, the policies would bring more general societal benefits and relieve pressures to tap fragile natural resources for fossil fuel. For example, reducing the combustion of coal, oil, and natural gas—by substituting renewable energy and by improving energy efficiency—would not only reduce emissions of carbon dioxide, mitigating the greenhouse threat, but would also bring other benefits:

- Burning less coal and petroleum would cut emissions of sulfur dioxide and nitrogen oxides, the chief precursors of acid rain and smog.
- Fossil-fuel emissions are also precursors of tropospheric ozone, another greenhouse gas, which is also toxic. It impairs human health and damages forests and crops.

- Reducing our burning of fossil fuels would also reduce the amount of carbon monoxide in the air. This gas is also a poison and the principal reason some cities, notably Denver, do not meet federal clean air standards. Carbon monoxide, incidentally, also contributes to the greenhouse problem: it destroys natural chemicals in the atmosphere that break down methane, and thus enhances the buildup of this greenhouse gas.
- Weaning ourselves from petroleum would reduce our dependence on foreign oil, mitigating all of the national security concerns related to that dependency.
- Energy efficiency—the most desirable and cheapest path to the goals outlined above—makes good business sense. According to calculations by Amory Lovins, the founder of the Rocky Mountain Institute and a leading scholar of energy use, if the United States became as energy-efficient as Japan and the nations of Western Europe, we would save something on the order of $200 billion a year.

Another high-leverage greenhouse policy is completely eliminating the use of chlorofluorocarbons. Chlorofluorocarbons accounted for about 24 percent of the human-induced greenhouse effect during the 1980s and, if not controlled, could cause more than one-quarter of the additional warming expected over the next 100 years. These powerful greenhouse gases are also the principal reason the planet's stratospheric ozone layer is eroding, and with it our protection from incoming ultraviolet radiation. Eliminating these chemicals would make a huge difference to two of the most serious global environmental threats.

Similarly, reducing the rate at which we are destroying tropical forests, a practice that currently releases 1 to 2 billion tons of carbon into the atmosphere every year, would bring benefits beyond slowing down the rate of greenhouse warming. Presently we are destroying 100 acres of forest every minute of every day; those forests contain somewhere between 50 percent and

90 percent of all species of plants and animals on Earth. We are pushing many of those species into extinction even before we know what they are, or have learned what new medicines, new crops, or new materials they may provide. Preserving tropical forests will also maintain a crucial sink for carbon dioxide. The total amount of carbon stored in the forests of the tropics is about a third of that in the atmosphere, so the amount that stands on the brink of release—subject to the whims of developers and slash-and-burn agricultural practices—is huge. Planting trees anywhere can increase the planet's carbon dioxide sink, and would reap the additional benefits of cooling urban areas in the summertime and reducing the need for air-conditioning.

Finally, extracting fossil fuels for energy threatens natural ecosystems, such as Alaska's Arctic National Wildlife Refuge. Using less fossil fuel would decrease the pressure to open up our coasts and wilderness areas to drilling. Fuel transportation and distribution runs the risk of both small and large accidents that destroy wildlife and habitat. The 1989 *Exxon Valdez* spill in Prince William Sound is only one example.

Some have labeled this a no-regrets strategy. No matter what happens to the global climate, the steps required to control greenhouse gases will lead to improvements we'll welcome anyway.

Even the White House subscribes to the notion that the greenhouse effect is worth dealing with if mitigation measures are worth taking for other reasons. In 1991 the Administration proposed a fourfold approach. It includes:

- A commitment to phase out the manufacture of chlorofluorocarbons by the year 2000, mainly to protect the ozone layer. An additional benefit, of course, would be controlling the greenhouse effect of CFCs.
- Support for the acid-rain provisions of the Clean Air Act passed in October 1990. Those provisions, by inducing utilities to wean themselves off highly polluting (and green-

house gas–emitting) coal, are comparable to putting 22 million automobiles (one-fifth of the American fleet) in mothballs for the next decade.
- A commitment to cut energy use in government buildings by 20 percent by the year 2000 and to cut energy use in government cars by 10 percent by 1995.
- A presidential proposal for a combined public-private initiative to plant 1 billion trees a year for five years on private land in this country. Those trees could eventually soak up 13 million tons of carbon annually (about 5 percent of the country's current carbon emissions) if the program continued for 20 years rather than five.

But these policies are inadequate to address global warming. As the National Academy of Sciences (NAS) said in its 1991 report, steps should be taken to reduce greenhouse gas warming regardless of other benefits. Emissions of greenhouse gases must be cut substantially in order to slow the impending warming to a rate to which society can adapt. The world's richest and most energy-consuming nation must demonstrate how. The goals are clear. Carbon dioxide emissions must be cut by increasing energy efficiency and substituting energy sources that do not emit carbon dioxide. Other greenhouse gases must be controlled. The deforestation trend must be reversed by halting the destruction of tropical and other forests and by initiating a massive program of replanting.

Can this be done without undercutting economic growth and without spending more to decrease greenhouse gas emissions than such reductions are worth? By one estimate, current technologies could reduce worldwide emissions of carbon dioxide 25 percent within 15 years. If this achievement was transcended by eventually reducing carbon dioxide by more than half (and by controlling other greenhouse gases), the average warming rate could be kept to .1 degree Celsius (.18 degree Fahrenheit) per decade over the next century.

William Nordhaus, an economist at Yale University and former member of the president's Council of Economic Advisors, estimates that greenhouse gas emissions could be reduced by 10 or 20 percent "without a noticeable effect on economic growth," and that even a 30 percent reduction could be achieved fairly painlessly. The NAS asserts that a 10 to 40 percent reduction could be achieved with little economic impact. Although actual costs can be debated, the overall message is hard to dispute: the world *can* afford to begin to tackle greenhouse gas emissions.

International Policies

Many other nations agree, and almost all Organization for Economic and Cooperative Development (OECD) countries have committed themselves to such greenhouse-fighting policies. In stark contrast to the United States, their strategies include numerical goals and dates for reductions or limitations in carbon dioxide. A sampling:

- Countries that are members of the European Community and the European Free Trade Association plan to stabilize carbon dioxide emissions at 1990 levels by the year 2000. Canada says it will achieve this stabilization "as a first step" to more stringent controls.
- Great Britain plans to stabilize carbon dioxide emissions at 1990 levels by 2005, provided other nations do their share.
- Germany plans a 30 percent reduction in carbon dioxide emissions from 1987 levels by 2005.
- Australia and Austria plan emissions cuts from 1988 levels of 20 percent by 2005; Denmark is planning to reduce its 1988 carbon dioxide emissions by 20 percent by the year 2000, and is considering halving emissions by 2030.
- The Netherlands will stabilize emissions by 1994, then implement cuts of 3 to 5 percent by 2000 and follow those

with more substantial reductions. New Zealand proposes to cut emissions 20 percent of 1990 levels by 2000.

It is particularly instructive to examine the greenhouse policies of Japan. As the world's rising economic powerhouse, it is particularly hostile to measures that would undercut its productivity. But as the recent target of much environmental criticism—for its whaling, its logging, its drift-net fishing—Japan has also become acutely sensitive to its image on the world stage. In late 1990 the government adopted a plan to either hold per capita emissions of carbon dioxide at 1990 levels by the year 2000 (a policy that would allow total emissions to grow, since population is expected to rise 5 percent in the decade) or hold actual emissions to 1990 levels. Levels of methane, nitrous oxides, and other greenhouse gases would be held to their 1990 levels "as far as possible."

To achieve either goal, the Japanese government will expand green areas in the cities to absorb carbon dioxide and will encourage the use of insulation, solar heating, and generating electricity from waste heat in order to burn less fossil fuel. It will promote greater use of mass transit, further develop nuclear power, and introduce small-scale power sources such as photovoltaic cells that turn sunshine into electricity. These measures, too, would decrease reliance on fossil fuels. The public will be encouraged to recycle even more than it does now; overpackaging and direct mail—junk mail—will be discouraged in order to reduce the demand for new paper and other materials whose production eats up energy.

The Japanese plan is not perfect. First, it relies on expanding nuclear power; second, the new measures are not legally binding, but then, that may not be so serious in a nation where consensus provides a powerful force in shaping behavior. The main point is that Japan has set a greenhouse-gas goal and a timetable, and has outlined the strategy for achieving it.

There is one other irony. Japan is not the worst greenhouse

gas culprit. Its per capita emissions of carbon dioxide amount to about 2.1 metric tons per year. The comparable figure for Great Britain is 2.8 tons per capita; for what was formerly West Germany, 3.0 tons. And the United States? A scale-tipping 5.0 tons. Clearly this country has some way to go in cutting its greenhouse gas production. Other industrialized nations believe that this can be achieved without crippling their economies.

Goals for the United States

In order to seize the initiative, the United States must clean its own house without waiting for more recalcitrant countries to come on board. There are two reasons for putting America on the spot. First, we are the world's leading contributor to the greenhouse problem in total emissions and thus the obvious leader to show the rest of the world the way out of the greenhouse (see Figure 5). Second, a definitive greenhouse policy in the United States would spur the development of technologies to achieve reductions in greenhouse gases everywhere.

That is more than a matter of global altruism. Japan has already established a $40 million research institute to develop environmentally safe technologies, including those to combat greenhouse warming. Sixty major companies—from NEC, Sharp, and Hitachi to Nippon Steel, Toyota, and Nissan—are contributing money and employees. There is money to be made in antigreenhouse research and development. If only for reasons of international economic competitiveness, the United States would do well to join the search for carbon-dioxide-reducing technologies. Otherwise, we will end up purchasing technologies from other, more farsighted nations. The sooner the United States commits itself to the importance of those technologies, the less likely are we to find ourselves buying them from abroad.

A word about the emphasis on government actions and policies: With the twentieth anniversary of Earth Day in April 1990, it became popular to focus on things individuals can do to save the Earth. It is all well and good to buy the compact gas-efficient car, insist on organic produce, and recycle at every opportunity. But individual actions will have an impact only insofar as government and corporate policies create a context in which individual actions have meaning. What good does it do to try to recycle one's air conditioner's chlorofluorocarbon coolant if no repair shop nearby offers that service? There must be government incentives to establish such services or to create such a demand for them that market forces bring them about.

A variety of policy actions need to be pursued by the federal government in order to achieve substantial greenhouse gas cuts. Obstacles that prevent market forces from operating (such as subsidies that support inefficient uses of energy) should be removed, and concurrently, taxes that appropriately account for the full environmental costs of greenhouse gas emissions (such as a tax on gas-guzzler vehicles) could be enacted. In addition, innovative systems for cutting emissions, such as trading permits or allowances to emit greenhouse gas on an international level, could be implemented. Last, the federal government could greatly boost market demand for energy-efficient technologies through its procurement policies, by purchasing only highly efficient autos for its fleet, for example. All four types of policies should work together to provide incentives and constraints that reduce greenhouse gas emissions.

Reducing Carbon Dioxide Emissions

The Intergovernmental Panel on Climate Change estimates that in order to stabilize the atmospheric concentration of carbon dioxide at its current level, global emissions of carbon dioxide would have to be slashed by 60 percent or more. Methane emissions would have to be reduced 15 to 20 percent, chlorofluorocarbons 70 to 85 percent, and nitrous oxide 70 to 80 percent. Clearly our work is cut out for us. As a down payment on those reductions, scientists and diplomats meeting in Toronto in the summer of 1988 to address "The Changing Atmosphere" called for a 20 percent decrease in carbon dioxide emissions by 2005.

To reach either of these two emission-reduction goals, it will not be enough for the world's developed countries, which use about one-half of the world's energy, to cut their consumption. The rate of commercial energy consumption in developing countries (such as India) and centrally planned countries (such as China) is growing even faster than in the developed countries. Recent experience demonstrates that economic growth can be compatible with reduced energy demand. For example, by implementing efficiency measures, the United States has managed to keep energy use at a little over 1973 levels, even as its economy grew by 40 percent.

Viewed another way, between 1972 and 1985, the amount of energy America used to produce one dollar's worth of output fell by more than 25 percent. Nearly two-thirds of this was due to improvements in energy efficiency, not only in direct energy use (such as using less electricity to run an engine) but also indirectly, as by substituting less-energy-intensive materials such as plastics for more-energy-intensive ones such as steel. The rest of the energy gain came from the gradual disappearance of energy-intensive manufacturing and the surge in less-energy-intensive service industries such as finance and health care.

Despite such gains, the United States still uses far more energy to produce a unit of gross domestic product (GDP) than many other developed nations. This ratio of energy to GDP is called energy intensity. Our energy intensity is twice that of western Germany, Japan, Switzerland, and Denmark. This profligate use of energy explains in part why the United States accounts for one-fourth of global energy use, despite having only 5 percent of the world's population.

Reducing emission of carbon dioxide substantially will require both energy efficiency—a solution effective in the near term—and increased reliance on alternative means of energy

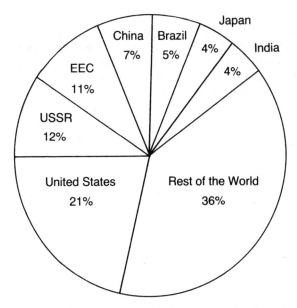

FIGURE 5. *Contributions to global warming in the 1980s by country (percent)*

production, a solution that will take decades to implement fully.

Nearly every sector of the economy—transportation, manufacturing, electricity generation—offers substantial opportunities for improved energy efficiency. The following policies and measures describe those opportunities.

Fuel Economy Standards

Transportation accounts for a little over 25 percent of the country's energy use. Nearly 80 percent is used by cars, light trucks, and freight trucks, the rest by aircraft, buses, trains, and other vehicles carrying more than a few passengers. The United States passenger car fleet, at 137 million of the approximately 400 million such vehicles on the planet, is the largest in the world; we have one car for nearly every other man, woman, and child in the country.

Passenger cars are responsible for about 13 percent of the annual emissions of carbon dioxide in this country and the same percentage of total carbon dioxide emissions from fossil fuels worldwide. That figure is expected to soar 75 percent by the year 2010.

Emissions are directly proportional to fuel economy. The farther a car goes on a gallon of gas, the less carbon dioxide it emits per mile. A car that gets 20 miles per gallon emits 7.5 tons of carbon dioxide over the course of 15,000 miles (the average number of miles driven in a year in the United States), while a vehicle that gets 50 miles per gallon emits only 3 tons driving the same distance. In the United States, if CAFE (Corporate Average Fuel Economy) standards do not rise, and if the current rate of growth in vehicle miles traveled continues, carbon dioxide emissions from the nation's transportation sector will increase some 40 percent by the year 2005. Thus the first step toward reducing the amount of carbon dioxide coming from vehicles is to increase the fleet's fuel economy.

There has already been considerable progress. New cars spew out, on average, 2,000 fewer pounds of carbon every year than they did in 1974. The 1990 fuel economy standard for new vehicles is 27.5 miles per gallon. It is technologically feasible to increase that considerably. Some models, for example the Honda Civic and Geo Metro, already exceed 50 miles per gallon. If all vehicles on the road today averaged 50 miles per gallon, we would use 40 percent less gasoline (cutting our oil imports more than 10 percent) and our vehicles would be responsible for approximately 300 million fewer tons of carbon dioxide every year. The American Council for an Energy Efficient Economy, a nonprofit research group, estimates that a high fuel standard also makes good sense for a buyer's pocketbook. Using a benchmark of 40 mpg, the Council concludes that the extra $500 price tag for engine and other modifications would be more than offset by $2,000 in gasoline savings over the life of the car. According to the Council, a 40 mpg standard would also keep 124 million tons of carbon out of the air every year. Unfortunately, the United States virtually abandoned fuel efficiency research and development in the mid-1980s, when the real price of gasoline plummeted. Meanwhile Europe and Japan, where gasoline costs as much as triple what it does in this country, have the lead. There is no reason why Americans should not be driving cars that get 100 mpg in the very near future, and that standard should be the goal we set for ourselves.

Contrary to the warnings of the big auto-makers, cars can be both fuel efficient and safe. There are new materials that can make cars lighter without sacrificing comfort or safety. In addition, fuel economy can be boosted by using engines with four valves per cylinder, rather than the standard two; this is the tack taken by some GM, Honda, and Toyota models, for instance. A continuously variable transmission, as on the Subaru Justy, also increases mileage. Simply making the car more aerodynamically sleek improves its gasoline mileage by reducing drag. Electronically controlled automatic transmissions helped improve miles per gallon by 10 percent in Toyota and Mitsubishi models. The Department of Energy estimated in 1989 that the fuel efficiency of new cars could rise 40 percent merely through the adoption of technologies already in use, without any breakthroughs or downsizing. Several makers have prototypes that get 65 miles per gallon in city driving and 75 miles per gallon on the highway. Unfortunately, because of their high cost and long payback period, they are not yet marketed.

There are ways other than mandating a higher CAFE standard to achieve higher fuel efficiency in vehicles. One way is to revise the tax system to encourage the purchase of fuel-efficient cars. Currently, a gas-guzzler tax (established in 1978) exists to do just that. If we extend the current gas-guzzler tax to more cars with low fuel economy, revenues from this tax could be returned, in the form of rebates, to consumers who buy highly efficient cars and trucks.

Market Mechanisms

An innovative way to encourage fuel efficiency and decrease gasoline consumption is to limit carbon dioxide emissions and allow trading of carbon dioxide "credits." A first step in such an approach would be to limit carbon dioxide emissions from stationary sources (such as power plants), but allow auto manufacturers to obtain emissions reduction credits for building cars that improve on the legislated fuel-efficiency standard. They could sell those credits to stationary sources that are required to reduce emissions. A more comprehensive way would be to distribute a strictly limited number of carbon dioxide allowances to the main fossil fuel producers, such as the oil industry. The end result would be increased fuel costs, which in turn, would cause consumers to demand more fuel-efficient cars and cause alternate fuels to become more competitive.

Another way to discourage use of gasoline is to enact a carbon tax. Taxes on fossil fuels, such as a gasoline tax, that are set according to their greenhouse gas contribution, would encourage conservation and render explicit the true costs of fossil fuels. Such taxes would need to be high enough to cause individuals and industry to use alternative means of transportation or energy production. The revenues could be applied to building mass transit systems and to enhancing the development and use of renewable energy sources. Or they could be returned to the consumer by cutting other taxes.

It is difficult, however, to know how large to set the tax to get the sought-after environmental result. If, for example, the goal is a 20 percent reduction in gasoline use, it is not clear how large the tax should be. Another consideration is the length of time it takes for a tax to get results. Research indicates that it takes 10 to 15 years for consumers to fully respond to a gasoline tax. Research also indicates that over the long term, consumers will react to a tax by buying more fuel-efficient cars, rather than changing their driving patterns.

Transportation Funding Policies

In order to reduce carbon dioxide emissions from cars and trucks, however, gains in fuel efficiency won't be enough. They must be paired with reductions in the total number of miles traveled by the fleet. This total has been increasing at a rate of 2.5 percent nationally and 3.5 percent or more in most cities, with increases as high as 6 percent a year in some high-growth areas in the western United States. The total is expected to soar another 50 percent between now and 2010. If the current rate of increase continues, fuel economy would have to rise to more than 40 miles per gallon simply to negate the effects of those extra miles. Without a reduction in vehicle miles traveled, we will achieve virtually no net reduction in carbon dioxide emissions from cars and trucks. We will simply be running faster to stand still. For this reason, federal transportation funding policies must be revised to promote alternatives to the use of individual vehicles.

Eighty-five cents of each federal transportation dollar is currently dedicated to promoting vehicle use. The law requires that these dollars must be spent on highways and other infrastructures for moving cars, even if state or local leaders prefer to develop transportation systems that move people rather than vehicles. Federal transportation funding must be made flexible so that communities can build transportation systems that let people leave their cars at home. Almost anything short of a military tank is a better way of getting around. Fully utilized intercity rail and bus systems can use only one-tenth the amount of energy per passenger mile that the family roadster does; light rail, city bus, and rapid rail systems use one-seventh to one-sixth of the energy.

Minimizing Car Use

States and localities have a significant role to play in reducing the environmental impact of vehicles. If they plan local and regional transportation to minimize individual car use, it would help clean the air, too. One framework for transportation planning is the Southern California Association of Government's Regional Mobility Plan, which aims to reduce the growth of vehicle miles traveled by 60 percent by 2010. Vehicle use is expected to increase by 150 million miles per day in the four-county southern California area from 230 million to 380 million miles by 2010. Through a broad mix of strategies designed to provide alternatives to use of the individual auto, including rail transit, express bus lanes that will move bus travelers faster than those in cars, and programs to encourage shorter and fewer trips, southern California aims to reduce this increase by 90 million miles a day.

Local governments can also, through their regional planning policies, restructure cities and suburbs to make life without a car both more desirable and more practical. That means compact cities where homes, jobs, shopping, and other necessities are near public transit as they are in Europe. Sweden's cities, for instance, are typically compact hubs surrounded by great sweeps of rural, generally forested land. Stockholm is circled by satellite communities of 25,000 to 50,000, linked by rail and expressway. Homes, offices, and shops are concentrated around train stations, which connect people to their jobs, homes, and services. Paris has a similar design. Although it is easier to build such a city from scratch, it is also possible to change long-established land-use habits. Toronto offers developers incentives to build near subway stops, for example. Inducing builders to site residences in metropolitan centers can also counteract years of suburban sprawl and inner-city decay.

To entice drivers out of their cars and into buses and trains—or even onto bicycles—governments can also stimulate the use of mass transit and high-occupancy vehicles, as by setting aside high-occupancy vehicle lanes that make sure a vehicle carrying many people will get to its destination before the vehicle with only a driver. States and municipalities can enact taxes or fees to make the price of using roads, parking, and filling the gas tank more accurately reflect the environmental cost of vehicles. This can range from a higher state tax on gasoline (close to the $1 or $2 per gallon common in Europe) to elimination of the tax-free status of employer-subsidized parking and company cars. The more expensive it becomes to drive, the more people will press for improved public transit systems.

Alternatives to Gasoline

Just as making a gallon of gasoline go further controls carbon dioxide emissions somewhat, so does running cars and trucks on fuels other than petroleum, which now supplies 97 percent of the energy used in transportation. Burning a gallon of gasoline releases about 20 pounds of carbon dioxide; some of the alternatives are marginally or significantly better:

Compressed natural gas. At first glance, natural gas should produce fewer greenhouse emissions than gasoline: as a rule of thumb, oil contains 44 percent more carbon per unit of energy than natural gas, and coal contains 75 percent more. Natural gas is a proven technology: more than 500,000 compressed-natural vehicles ply the roads of the world today. One American company now offers a conversion service: for $2,500 to $3,500, Natural Fuels Corp. of Denver will convert a standard gasoline-burning car or minitruck into one able to burn either gasoline or natural gas—with the flick of a dashboard switch. In the conversion, natural gas cylinders are mounted in the rear of the vehicle and a natural gas line is connected to the regulator unit that feeds fuel into the engine. Natural Fuels's branded natural gas motor fuel costs 65 cents for a gallon-equivalent of gasoline, promising a payback for fleets in about three years, and for privately owned cars in about five. The firm plans to open such conversion centers elsewhere in Colorado, and in Texas (which sits on huge fields of natural gas), by 1992. But in 1990 the U.S. Office of Technology Assessment concluded that because natural gas currently produces more methane emissions than gasoline during production and transport, its overall greenhouse impact is only slightly less than that of gasoline-powered cars, if used in natural gas–dedicated vehicles. However, some of those methane emissions could be curtailed by tightening pipelines and reducing leaks.

Ethanol, or grain alcohol. In 1987, the United States produced 780 million gallons of ethanol, which is commonly combined with gasoline and marketed as gasohol. If 10 percent of the U.S. car fleet ran on such alternative fuels, the country would save 800,000 barrels of oil a day. The effects on greenhouse gas production are more speculative. They depend on the feedstock ethanol is made from, how efficiently that crop (currently corn) is grown, and how efficiently it is processed into ethanol. Using the most optimistic assumptions about efficiency, the U.S. Environmental Protection Agency concluded in 1990 that vehicles running on pure ethanol might account for 20 percent less carbon dioxide emission than vehicles running on gasoline. However, in addition to these optimistic assumptions, EPA did not consider ethanol's impact on other greenhouse gases, such as the nitrous oxide that would be generated by adding nitrogenous fertilizer to corn crops. Using more realistic assumptions, EDF concluded in 1991 that no net reduction in carbon dioxide would result from ethanol, and that an increase in other greenhouse gases could occur instead. The greenhouse benefit would be greater if the ethanol were derived from trees planted on marginal lands. The growing trees would absorb carbon dioxide, and then each year the wood from some of them would be turned into ethanol. Additionally, advanced processes for making ethanol from other materials such as wood could yield up to a 70 percent reduction in greenhouse gas emissions compared to gasoline.

Electric. There has been significant progress on electric vehicles. Again, technology is being driven by social policy. In order to control Los Angeles smog, the South Coast Air Quality Management District has mandated that, by the year 2003, 10 percent of the cars sold in the state must have zero emissions. That standard can be met only by electric vehicles. Because California is such a huge market, the major manufacturers will be forced to produce the vehicles in quantity and at an acceptable price. One manufacturer, Clean Air Transport, will offer 30,000 electric vehicles in the state by 1995. Solar Electric of Rohnert Park, California, offers vehicles ranging from three-wheel electric mopeds ($995; speed up to 15 miles an hour, range of 20 to 40 miles) to converted Fiat X-19s ($7,000; 60 mph; 40–45 miles) to converted Ford Aerostar vans ($28,500; 55 mph; 50–65 miles).

The biggest barrier to electric vehicles has been the storage battery. Traditionally a lead-acid battery, it is heavy and short lived, requiring replacement after every two years or so (at a cost of $1,500). Recent improvements in battery technology could make electric cars more feasible. General Motors has developed the sporty Impact, which can travel 120 miles on a single charge of its lead-acid battery and zoom from 0 to 60 in eight seconds. Recharging the battery, from standard 220-volt household current, takes between two and six hours. In 1990, GM began marketing the electric G-Van. Chrysler's TEVan, with a nickel-iron battery, is in development and scheduled to reach dealers by 1994; it will get 120 miles on a charge and reach 65 miles an hour. Ford is developing an electric van that runs on a sodium-sulfur battery.

Even if electric vehicles get their charge from the average fuel mix expected to be used by electric utilities in the year 2000, greenhouse emissions per mile driven would be 25 percent less than from gasoline. If the charge comes from a utility that burns natural gas, the car would be responsible for 36 percent less greenhouse gas emissions.

Charging the batteries with electricity produced by solar or other renewable sources would completely eliminate greenhouse gas emissions attributable to driving that car. Even though an electric vehicle powered by coal-derived electricity would not produce any fewer greenhouse gas emissions, the reduction in air pollutants, such as hydrocarbons and nitrogen oxides, would be environmentally beneficial.

Hydrogen. Further from widespread introduction than electric vehicles, hydrogen-powered vehicles are the subject of intense research. The fuel would have to be stored in clunky vessels. A current prototype has a storage vessel that weighs about 800 pounds and can hold enough hydrogen to propel a typical vehicle only about 75 miles. Still, if the hydrogen were produced by the electrolysis of water—as in the standard high school chemistry experiment—and if the electricity for that electrolysis came from renewable sources, a hydrogen-powered vehicle would be responsible for no carbon dioxide emissions. Although its fueling system poses complications, it is certainly an interesting and worthwhile idea for the future.

Promoting Least-Cost Planning

In the utility sector, there is a great potential to increase efficiency through least-cost programs. The principle behind least-cost planning (LCP) is to rank all energy options, from conservation to new energy sources, on a scale of cost-effectiveness and then to use the least expensive means (calculated over the whole life cycle of the technology) of meeting energy demand. Conservation and efficiency measures usually prove more cost-effective than developing new sources of energy. Utilities need to use LCP approaches to help their customers conserve energy. For example, supplying customers with free compact fluorescent light bulbs is an effective way to demonstrate conservation and reduce energy demand. Forty-two states have developed, are developing, or are considering least-cost planning policies, but as of 1989 only 17 states had such plans in place or were in the process of adopting them. Promoting LCP through regulatory programs and market incentives is a prime area for state government action. There are four ways for states to promote LCP utility services: provide utilities with financial incentives to adopt least-cost plans and energy efficiency measures; allow energy-efficient investments to compete with energy supply options in competitive bidding by utilities when purchasing power; encourage least-cost planning through federal regulations of interstate power sales; and conduct research to improve utility, state, and local energy conservation programs.

Since states set utility rates, they should also make energy prices in the market incorporate the environmental costs of energy production. Those costs should reflect not only the contribution of a particular energy source to the greenhouse effect but also its contribution to other environmental degradation such as acid rain and smog. There are many ways to do this.

Improving Energy Efficiency in Industry

Industrial emissions of carbon dioxide account for about 32 percent of the total in the United States—a little more than transportation. The largest users of energy in the manufacturing sector are the paper, chemical, petroleum and coal, and primary metals industries. The United States and other industrial countries have lowered industrial energy use 30 percent since the oil shock of 1973. But despite gains in reducing energy demand in this sector, the United States is still less efficient than its major competitors; the United States has an industrial energy intensity twice as high as western Germany's. The three main areas of industrial electricity demand—motors, electrolysis, and electric resistance heating—are the ripest areas for increasing industrial energy efficiency in the United States.

The Electric Power Research Institute estimates that 24 to 38 percent of industrial energy use could be saved by the year 2000. Motor drives, which turn fans and blowers, run pumps and compressors, propel conveyor belts and process lines, offer the biggest opportunity. They account for 67 percent of the electricity used by industry. Motors with adjustable speed drives could slice industry's use by 20 percent, EPRI concludes. Additionally, energy management systems (those that can turn off equipment when appropriate) and improving the energy efficiency of operations, can help curb carbon dioxide emissions.

Encouraging Energy Efficiency in Buildings

Homes and office buildings account for a little more of the total carbon dioxide emissions in this country than transportation does (36 percent), and they too offer a large potential for energy efficiency. Most of the carbon dioxide emissions from buildings are actually indirect emissions: they come from our use of electricity, the generation of which by fossil fuels spews out carbon dioxide. Space heating accounts for 43 percent of building emissions, appliances for 20 percent, air-conditioning and lighting for 14 percent each, and water heating for 9 percent. Thus all of these areas—electric lights and appliances, heating, and cooling—offer opportunities for energy efficiency and consequently for reductions in the emission of carbon dioxide.

Insulation. A financial payback makes the environmental benefits look even better. Buildings can be designed and constructed to cut fuel bills without adding to their capital costs. Construction costs for energy-efficient commercial buildings are barely more than for conventional buildings and may be less, thanks to savings from purchasing smaller heating and cooling units. In prototype "superinsulated" homes, the energy required for heating and cooling is a mere 10 to 25 percent of the average in American homes. These homes have well-insulated walls and ceilings and often include ventilation systems that recover heat from exhaust air.

Retrofitting existing buildings also saves money while saving energy. This is crucial to any greenhouse policy, for the country's existing buildings will be replaced much too slowly to rely on new construction as the only means to energy conservation. Arthur Rosenfeld, an energy efficiency expert at the Lawrence Berkeley Laboratory, estimates that for an investment of $2,000 to $7,000 to superinsulate a house, a homeowner could look forward to heating bills of a paltry $20 to $300 per year—even in frigid climes like Minnesota's. That's because these heat-tight homes store the warmth given off by people, appliances, and sunlight coming through windows.

Some of the most dramatic advances in home energy use have come in the Pacific Northwest, partly because of more stringent building efficiency codes. The electricity used to heat and cool well-insulated new homes in that region is only about 3.3 kilowatt-hours per square foot of space per year. More typical homes in the region draw 8 kilowatt-hours per square foot per year. In cold climates, buildings with double the standard amount of insulation and an airtight liner inside their walls use so little energy compared to conventional buildings that they pay back their added construction cost in about five years.

Lighting. Twenty to 25 percent of the electricity sold in America goes for lighting, so that use offers a golden opportunity for energy conservation and hence greenhouse gas reduction. Lighting accounts for about 10 percent of carbon dioxide emissions in this country. The simplest change is to replace incandescent bulbs with compact fluorescent ones that screw in just like incandescents. They draw between 60 to 75 percent fewer watts to provide the same amount of illumination as standard incandescents and last about 10 times longer, but can cost 20 times as much. That combination means that the payback period—the amount of time needed for the savings in electricity or fuel of an investment in energy efficiency to add up to the original cost of that investment—is about a year. After that, the investment has paid for itself and all future savings can be counted as profit. Each high-efficiency compact fluorescent bulb can reduce carbon dioxide emissions by 220 to 382 pounds a year (depending on the fuel the utility uses to generate electricity).

Even incandescent lamps can demand less energy: film on the inside of the lamp walls that reflects heat onto the filament while allowing visible light to pass through heats the filament, producing more light. A 60-watt coated incandescent lamp provides about as much illumination as a 150-watt ordinary incandescent. Occupancy sensors, especially useful in commercial buildings, can turn off lights when no one is around. These sensors use infrared or ultrasonic signals to detect movement. Photocells can sense the amount of sunshine coming into a room and adjust the lights to maintain just enough illumination. Finally, more efficient light bulbs throw off less heat, for instance, and thus reduce the amount of air-conditioning required in summer months.

Windows. Windows are another good place to seek energy savings. In the wintertime, they leak about one-third of the heat used in American homes—equal to about the annual output of the Alaska oil pipeline. Simply replacing ordinary single-glazed windows with double-glazed panes halves the energy loss. Coating one of the inner surfaces with a thin film of transparent "low-emissivity" material reduces heat loss to one-third what it was with single-glazed panes; these coatings, which include substances such as tin oxide, reflect heat back into the house. The next-generation improvement in windows will be to replace the air between panes with a gas that insulates better; xenon and argon are both better than air. Lawrence Berkeley Laboratory researchers have done even better: a window with a third layer of glass, two low-emissivity coatings, and krypton-argon filling. This one leaks a mere one-eighth what the inefficient single-glazed pane did, and even facing north it collects more heat on a winter's day than it loses at night.

Appliances. Appliances can also be designed to draw less electricity. A typical refrigerator run by electricity generated by burning coal causes the emission of 700 pounds of carbon a year. Efficient home refrigerators cost an extra $70 and use 40 percent less electricity than the average model in use today, thanks to better insulation and more efficient compressors.

Technologies are readily available to make other appliances run on less energy, too. Many electric water heaters heat through electrical resistance; gas water heaters and heat-pump water heaters are at least twice as efficient. The energy consumed by a conventional clothes dryer could be slashed 70 percent if the unit were replaced by one operated by heat pump. Induction cooking surfaces—which employ an electric field to generate a current in the pot or pan, producing heat—use less energy than both gas and electric burners.

The trouble is that there are several barriers to serious investments in efficiency. To take one simple example, a landlord whose tenants pay their own electric and heating bills has little incentive to install efficient windows, lighting, or appliances. His tenant, not he, will reap the benefits of lower utility bills while he will be stuck with a significant capital outlay. Even homeowners are reluctant to junk their perfectly adequate refrigerator for one that uses less electricity if that investment is likely to reward future occupants. As a result, studies have shown, individuals and businesses hesitate to invest in energy-saving devices unless they are confident of recovering the cost in no more than three years.

Energy Efficiency Standards. Since even that payback period is too long for many consumers, state governments can play a role in helping investors over the payback hurdle. One way is to mandate certain standards of energy efficiency for lighting, appliances, and construction. Such codes already exist at both the national and state levels and have been enormously effective in saving energy. In California, for instance, the energy efficiency codes already enacted will save the state about 13.5 billion watts of electricity by 2007 (7.8 gigawatts of that from construction standards and 5.8 gigawatts from appliance standards). The National Appliance Energy Conservation Act of 1987, which targets only appliances, will save the country some 13 gigawatts of capacity by 2000.

But the law could do more if it was aggressively implemented. It requires the Department of Energy to review and upgrade the standards regularly and to establish standards for other types of appliances. Despite clear evidence that standards reduce energy costs for consumers by billions of dollars while significantly cutting greenhouse gas emissions, DOE has seemed reluctant to fulfill its mandate.

To promote energy conservation, the federal government also needs to set minimum efficiency standards for all lights and appliances. Such standards have helped make the average refrigerator bought today more efficient than the best refrigerator available 10 years ago. Similar improvements could be made with lighting, which uses 20 to 25 percent of all electricity consumed in the United States. Minimum efficiency standards for fluorescent ballasts were set by Congress in 1988. But federal standards for fluorescent and incandescent lamps are still nonexistent. For incandescent lamps, the standard should be set to the level of today's krypton-filled lamps. Energy-efficient product labeling (similar to labels required for home appliances) for all lamps would assist consumers in making educated decisions regarding their lighting purchases.

Building codes. Building codes can ensure that energy efficiency is taken into account in new structures. California has adopted strict energy efficiency requirements, which between 1977 and 1985 saved 11 quadrillion Btus of electricity and 32 quadrillion Btus in gas. Another way to encourage efficiency is for utilities to reward building owners for efficiency and charge them for high energy consumption. This practice has been tested in Texas by the Rocky Mountain Institute. The federal government could provide substantial technical support to states and municipalities in developing codes through its experience in developing the Federal Building Energy Performance Standards.

Recycling. A simple way of reducing the amount of energy consumed, and hence total carbon dioxide emitted, by industry is to make fewer demands on our factories. In other words, require fewer products made from raw materials and more products made from recycled ones. The energy savings are significant. Making aluminum products from used aluminum consumes no more than 10 percent of the energy required to make the item from virgin alumina. Aluminum recycling can get a huge boost from state mandatory can deposit laws, such as those in New York, Oregon, Michigan, Connecticut, Maine, Vermont, and Massachusetts. Recycling steel saves between 45 and 75 percent of the energy otherwise required to fabricate steel from iron ore (the range reflects the varying efficiency of furnaces that melt the recycled steel). If states required consumers and business to recover steel scrap, it would encourage the construction of steel minimills with efficient electric arc furnaces to turn the scrap into saleable products again. Melting waste glass requires only 69 percent of the energy that melting virgin raw materials does. To encourage glass recycling, states can require people to separate their glass from other garbage and institute collection systems by municipalities so glass recyclers are assured a reliable stream of waste glass for processing. Recycling plastics can save between 92 percent and 98 percent of the energy otherwise needed to fabricate plastics from resins. Finally, recycling saves energy needed to extract and refine raw materials.

To encourage recycling, states, which have primary responsibility for managing municipal solid waste, should make it available to everyone. More than 14 percent of all municipal solid waste in the United States is currently recycled; the EPA's goals are that 25 percent should be recycled by 1992 and 50 percent by 1997. But recycling programs can't possibly meet these goals if they are not adequately funded. If states and cities funded recycling on a level with other forms of waste disposal, such as landfilling and incineration, the potential to recycle a significant portion of our garbage could become a reality. States need to use their procurement and economic development programs simultaneously to help expand markets for recycled materials.

Using Renewable Energy Sources

Another way to control the emissions of carbon dioxide from our factories and our utilities is to replace the fossil fuels used to generate electricity with renewable energy sources. Energy efficiency only slows the rate of greenhouse warming; replacing fossil fuels with energy supplies that yield few or no greenhouse gases would have a far greater impact.

The low- or nonpolluting alternatives to fossil fuel include energy sources known collectively as renewables. Unlike fields of oil, seams of coal, and pockets of natural gas, these fuels can be replenished about as quickly as they are tapped. They include solar thermal power and photovoltaics, biomass, and wind and hydropower. Ultimately many are derived from the Sun. The Sun makes living things grow that are burned as biomass, and it drives the air currents that turn windmills. Alone, none of the alternatives will replace the nation's use of fossil fuels, but together, they can be an important part of national and international programs to reduce emissions of carbon dioxide. As methods for renewable resource storage and transmission are developed, we can eventually bring about an end to fossil-fuel dependence. Here we briefly outline the technological and economic outlook for several renewables.

Biomass. This source includes burning wood and crop residues, methane derived from animal wastes, and "energy" crops, such as ethanol from corn. As long as the material burned is replaced with new plantings, there won't be any net gain of carbon dioxide in the atmosphere. Biomass currently provides 15 percent of the world's total energy, largely in rural areas of Asia. In the United States, the greatest biomass energy potential is from forest and timber wastes. If all available and recoverable energy potential was used, biomass could provide for 5.2 percent of the energy requirements of the United States.

Solar. Solar energy falls into four main categories: solar heating and cooling systems, commonly used to heat air or water; solar thermal technologies, which use collected heat to generate electricity; photovoltaics, which transform solar radiation directly into electricity; and solar ponds, which capture and store solar radiation by using the sun's energy to heat a mass of water or other liquid.

Solar thermal systems for low-temperature applications are usually characterized as passive or active. Active systems use force to move a heat transfer medium (usually air or water), while passive systems rely on convective forces. Collectors convert the Sun's radiant energy to heat, which is then used to heat water and heat or cool space.

As of 1987, there were over 200,000 fully passive solar buildings and approximately 1.2 million active solar installations. Sixty-four percent are used for heating water, 22 percent for pool heating, and 14 percent for space heating. The market for this technology could improve with appropriate incentives, such as tax credits. Solar water-heating systems are often competitive with electric water heating, such as passive solar buildings.

Large-scale solar thermal technologies concentrate energy to heat a liquid, which turns a turbine. These are appropriate for industrial processes and electrical generation. In 1990 the cost of solar thermal was only one-third what it was in 1985—as little as 8 cents per kilowatt-hour in sunny locations—and could drop by another 30 percent in the next few years. This form of solar power is also attractive as a peak-power source. One of the largest installations is in California. There, Luz International operates solar thermal power plants in the Mojave Desert, supplying power at 11.8 cents per kilowatt-hour. Luz expects that, as more plants come on line, the cost will drop, making solar thermal more competitive.

Photovoltaics hold great promise for growth. Despite little federal R&D support, their cost has come down remarkably in recent years. Even without a government push, their cost, now about 25 to 30 cents per kilowatt-hour, is expected to drop to 15 cents in 2000 and 6 cents in 2020 for high sun regions. But a little R&D investment can go a long way. The Solar Energy Research Institute estimated in 1990 that with another $36 million a year invested in research and development, the cost of electricity from photovoltaic cells could drop to 15 cents per kilowatt-hour by 1995 and 10 cents or lower by the year 2000. Some of the more optimistic projections of costs for PV go as low as 2 to 3 cents per kilowatt hour.

Current prices still compare poorly to running a new coal plant at 6 cents per kilowatt-hour. Nevertheless, PV electricity can still find a niche in the energy mix. It is particularly promising for remote uses—off the power grid—such as rural residences, water pumping, even street lighting. The greatest hope is that by the year 2000 the cost of photovoltaics will drop enough to make them competitive for peak-power use, that is, the extra juice a utility must supply when air-conditioning or other demand soars (PV is already competitive under optimal conditions). It is not economical for a utility to install enough base generating capacity to meet its highest-ever demand. When customers require more than the utility can provide, the company may purchase power from other utilities. Or it can rely on photovoltaics in sunny climates.

Solar ponds use water as a direct absorber of solar radiation and as a storage medium for the collected thermal energy. Salinity Gradient Solar Ponds (SGSP) are the most advanced; they mimic naturally occurring, highly saline lakes that have very warm bottom water. They require continuous insulation, access to water and salt, and large land area. SGSP costs consist largely of construction and instrumentation expenses. Electricity costs range between 15 cents to 35 cents per kwh. Other types of solar ponds include membrane-viscosity-stabilized, gel, immiscible, and shallow-water.

Wind. Between 1981 and 1985, the amount of wind power generated around the world increased seventeenfold. Currently the cost of wind energy ranges between 6 cents and 10 cents per kilowatt-hour at an average site in the Great Plains, and between 6 cents and 8 cents per kilowatt-hour at a California wind farm. The American market is expected to decline in importance relative to the world market over the next decade, partly because of the elimination of federal energy tax credits in 1985 and partly because of plans in other countries to increase investments in wind energy. Europe, for instance, is expected to install between 200 and 4,000 megawatts of wind power during the 1990s.

The United States need not be left behind. Wind power now supplies more than 1 percent of California's electricity needs, for instance, through more than 15,000 wind turbines installed mostly in three blustery mountain passes. Their aggregate power rating exceeds 1,500 megawatts. In Hawaii, 23 megawatts of wind-driven turbines generate commercial electricity. Wind is a proven technology, in addition to being virtually nonpolluting and renewable.

It is also an improving technology. The rotors, derived from helicopter design, have improved, as have the turbines, which have become larger and higher rated, thus improving the economics of wind power. Pacific Gas and Electric of San Francisco, the nation's largest electric utility, has joined forces with the Electric Power Research Institute (EPRI), an industry R&D center, and with U.S. Windpower, the largest American windpower firm, to develop a large turbine suitable for utilities during the 1990s. As luck would have it, the California sites where the wind farms are installed dovetail perfectly with the pattern of electricity demand: winds blow stronger in the summer and, at one site, in the late afternoon—just the season and time of day when electricity demand peaks. A recent study by EPRI and U.S. Windpower concluded that it is possible to produce a utility-class wind turbine that would generate electricity for 5 cents per kilowatt-hour or less. The president of Windpower, Dale Osborn, says 3 cents per kilowatt-hour "is not out of reach." In fact, the economics of wind is now so promising that tax credits are no longer required to lure investment in the technology: since state tax credits expired in California in 1985, 622 megawatts of wind power have been installed.

Although 90 percent of the country's wind power is in California, most of the best places for such capacity are someplace else. The Bonneville Power Administration, for instance, has identified sites in Idaho and Montana that could sustain between 20 and 40 megawatts of capacity.

Renewable energy now accounts for about 13 percent of the nation's electricity generation and 10 percent of the energy supply. Biomass accounts for half of that, hydropower 45 percent, and solar, wind, and geothermal 5 percent. The technologies are far from mature and hence could easily benefit from an infusion of dollars and talent. Technological improvements over the past decade have cut the cost of solar- and wind-generated electricity by between 60 and 75 percent, for instance; similar cost reductions can be expected over the next few years if research receives adequate support.

Research and Development

Federal research and development programs are needed in many areas, including renewable energy sources, conservation, efficient vehicles and appliances, energy-efficient industrial processes, demonstration projects, technology transfer, and information dissemination. Between fiscal year 1981 and fiscal year 1988, budget appropriations for renewable energy R&D dropped from $900 million to under $200 million. As of 1986, the United States ranked lowest among OECD countries in percentage of gross national product (GNP) spent on this area.

Beyond research, governments at any level can spur the introduction of renewable energy sources by helping to reduce the (often high) initial capital costs of the technologies. This can be accomplished through low-interest loans, loan guarantees, and tax credits.

Nuclear Power

Nuclear power is another alternative energy source, though not strictly a renewable one because uranium fuel gets used up in standard fission reactors. The rise of concern about the greenhouse effect during the unusually hot, dry summer and midwestern drought of 1988 raised the possibility of a renaissance for the nuclear industry. Nuclear power is indeed an alternative to fossil-fueled energy, since nuclear plants do not release carbon dioxide into the atmosphere. But such major problems as safety, cost, and waste disposal cast doubt on its long-term acceptability as well as on its practicality, especially in the developing world. Fission power will remain part of the energy mix, but there are solid reasons why its use will not be suddenly expanded.

First of all, nuclear power would have to grow throughout the world, not only in the developing nations that have the most experience with and expertise in it. In order to seriously diminish greenhouse gas emissions, their growth in the developing world will have to be curtailed; otherwise, anything the richer nations accomplish will do little to slow atmospheric warming. To meet anticipated demand, hundreds of new nuclear plants would have to be brought on line in the developing world by 2010 (compared to a mere 23 in 1988).

How likely is this? Nuclear power requires more up-front capital than any other energy technology except, possibly, some gargantuan hydroelectric projects. For example, in 1988 the capital costs of building 5 new nuclear power plants in the United States ranged between $1.5 billion to $6 billion. With their current crippling load of debt, Third World countries are not in the best position to take on more financial obligations—even if international lending agencies, private banks, or aid programs were willing to extend the necessary credit. In addition, nuclear plants require a corps of highly trained engineers, scientists, and regulators to keep the plants running safely and to properly manage radioactive waste. So far, most Third World countries lack such home-grown expertise. Nuclear power, therefore, is not an easy solution to the complicated problem of weaning the world off fossil-fuel energy sources. Most likely, nuclear capacity will grow little in the future and provide no additional greenhouse benefit.

Reforestation

Besides decreasing sources of carbon dioxide, the world can control this greenhouse gas by increasing absorption of it. One of the best carbon dioxide sinks is vegetation. The U.S. Department of Agriculture's Conservation Reserve Program (CRP) pays farmers to plant trees or other vegetation on highly erodible agricultural land. This program was part of a Food Security Act (1985) mandate to reduce erosion and nonpoint source pollution and cut agricultural surpluses. The program also helps increase carbon dioxide absorption by planting more trees. EDF found that planting trees on CRP land is the most cost-effective of several options for utilities to offset carbon dioxide emissions from future power plants. Depending on the species used, approximately 11 to 33 million acres of trees would offset emissions from all new power plants planned for the United States between 1987 and 1996. An EDF study found, however, that the number of acres being converted to forest land was lagging behind congressional goals.

The United States can also use its economic clout to promote greenhouse-wise policies overseas. The United States contributes billions of dollars every year to Third World development projects, through large banks such as the World Bank, and the U.S. Agency for International Development. Much of that aid has led to increased greenhouse gas emissions by promoting deforestation for agriculture and construction of large dams and by subsidizing fossil-fuel-fired power plants. The U.S. government could use its voting power at the banks and its aid programs to make environmental improvements the centerpiece of development aid decisions and to direct more aid to programs that increase energy efficiency and expand renewable energy resources.

Specifically, financial support from the United States to international development banks could support reforestation programs that maintain biodiversity and ensure an increase in the global carbon dioxide sink. Reforestation projects can reduce soil erosion, increase the use of sustainable crops, protect biological diversity, and provide income to rural economies. Because reforestation programs sometimes are coupled with destructive logging or deforestation plans, their benefits are not always guaranteed. They need to be ensured before projects receive financial and political support.

Deforestation and other land conversion practices result in emission of roughly 10 to 30 percent of the carbon dioxide that fossil-fuel combustion produces. Globally, forest and woodland areas have been reduced by 15 percent since 1850. Temperate forests in Europe and North America were cleared over the past few hundred years of agriculture, accounting for part of the historical rise in atmospheric carbon dioxide concentration. Today the tropical forests are being burned at an annual rate equal to the land area of West Virginia for marginal grazing and cropland. Only about 10 percent of the cleared land is ever reforested. Reversing this trend would not only reduce an important carbon dioxide source, it would increase a critical carbon dioxide sink. By one calculation, reforestation costs about $400 per acre. To compensate for the current rate of deforestation would require replanting about 25 million to 40 million acres per year. The bill would come to $10 billion to $15 billion per year—a small price for eliminating a major cause of the buildup of carbon dioxide (and which amounts to approximately the cost of two nuclear power plants).

Urban tree planting programs in particular have a great deal of potential for offsetting carbon dioxide emissions. Urban trees are 15 times more effective than forest trees in reducing carbon dioxide emissions, because of the indirect effects they have in cooling urban "heat islands." Planting 100 million trees, a goal set by the American Forestry Association in its Global ReLeaf program, would save 40 billion kilowatt-hours of energy by reducing cooling and heating loads. That amounts to 18 billion tons of carbon dioxide saved on an annual basis. Interstate,

state, and local highway corridors are excellent places to plant large numbers of trees.

Reducing Other Greenhouse Gases

So far, all of the policies discussed have addressed ways of cutting emissions or increasing absorption of carbon dioxide, the chief greenhouse gas. But as noted in Part 1, carbon dioxide is not the only greenhouse culprit. Atmospheric concentration of some of the others is increasing even more quickly than the carbon dioxide level. The others are more potent greenhouse gases, molecule for molecule. Some of them can be controlled—even eliminated—much more easily than cuts in carbon dioxide emissions can be implemented. For these reasons, any greenhouse strategy must place equal emphasis on the other greenhouse gases.

Chlorofluorocarbons

We can score the most direct hit on global warming by aiming at chlorofluorocarbons (CFCs). These gases, better known for eating up the stratospheric ozone layer, account for 24 percent of the current greenhouse effect (compared to 55 percent for carbon dioxide). This is an extremely inviting target—since eliminating CFCs would also help protect the ozone layer—and an easy one to hit, too.

For these reasons, a worldwide phaseout of chlorofluorocarbons must be implemented as soon as possible. The 1987 Montreal Protocol on Substances That Deplete the Ozone Layer, an international agreement, was amended in mid-1990 to require a full phaseout of some CFCs and halons by the year 2000. This agreement, while extremely significant, still falls short of applying to all ozone-depleting chemicals. A *full* phaseout of the production and consumption of all CFCs and halons will be needed in order to halt stratospheric ozone depletion and fully control climate change.

Proposed CFC substitutes (hydrochlorofluorocarbons and fluorocarbons) can drastically reduce ozone depletion, but will still contribute to the greenhouse problem, since some of them will be significant infrared radiation absorbers. Therefore, substitutes that are not greenhouse gases are needed. This is certainly not impossible. First of all, policy can be a strong inducement to technological innovation: a market for something such as chlorofluorocarbon substitutes can push ideas off the drawing boards, into the laboratory, and onto the market. Before the invention of the catalytic converter for automobiles, for instance, the automotive industry insisted that tailpipe emissions controls meant to reduce air pollution would sound the economic death knell for Detroit. Chlorofluorocarbons have a similar history. When they were banned in this country as aerosol propellants—in deodorants, hair sprays, and the like—industry had possible substitutes for only one-third of the remaining applications of these gases. All were projected to cost much more.

But necessity can be a wonderful prod to invention. In 1982 the U.S. Environmental Protection Agency was able to identify available substitutes for only one-third of the uses to which chlorofluorocarbons were put. Now that governments have agreed to phase out use of all chlorofluorocarbons, the EPA estimates that eliminating them entirely by the year 2000 will cost less than the $3 billion that was predicted in 1988 for only a 50 percent reduction. The substitutes range from water-and-detergent solutions for cleaning electronic circuit boards (one of the products is even based on orange peels) to ammonia for refrigeration. Depending on which substitutes are ultimately chosen, they could bring an unexpected dividend: many of the new coolants are more energy-efficient than their chlorofluorocarbon predecessors. As a result, says EPA, bringing them on line could produce energy savings of $5 billion in the decade of the 1990s. Clearly, political will is a potent driving force for technological breakthroughs.

Until these safe substitutes reach the market, however, and

as long as CFCs are around, users and manufacturers of chlorofluorocarbons and related compounds need to implement better management practices, such as capturing and recycling them. Needless emission of these gases can be eliminated by detecting leaks, and by capturing and recycling the refrigerants whenever a unit is repaired or junked. Refrigerants are used in automotive air conditioners, commercial air conditioners, and retail refrigerators and freezers. As prices for CFC-based refrigerants rise, capturing and recycling them will become more cost-effective.

Recycling the CFCs in auto air conditioners is also crucial. This use accounts for 25 percent of all the CFCs used in this country. Hawaii, Oregon, and Vermont already require that service stations repairing auto air conditioners own and use refrigerant recycling equipment within the next several years, and other states have similar legislation pending. States should also investigate incentives to promote off-site recycling centers run by refrigerant manufacturers or solvent reclaimers for large quantities of captured refrigerant from auto and commercial air conditioners and retail food refrigerators.

National refrigerant recycling regulations, which are currently being developed by the EPA, will soon require that all CFC coolants used in air-conditioning and refrigeration systems be recovered and recycled.

Methane

Methane, although more a product of nature than man, can largely be controlled. It is formed when bacteria decompose organic matter under anaerobic (no oxygen) conditions. If oxygen were present, the bacteria would produce carbon dioxide instead. Wet environments, such as swamps and rice paddies, emit methane, as do municipal landfills and the stomachs of ruminant animals (cattle, buffalo, goats, sheep), where anaerobic bacteria break down organic matter and produce methane.

Cutting back methane emissions promises to be difficult, because methane sources are numerous and diffuse. Nevertheless, even a small cut, about 15 to 20 percent, will bring atmospheric methane levels back into balance. One important contributor to methane emission, possibly ranking after rice production and domestic ruminant animals, is the inadvertent release of methane during the extraction, refining, and distribution of coal, petroleum, and natural gas. Coal-bed methane recovery and the replacement of leaky, outdated natural gas distribution lines would help minimize methane losses. Outside the United States, gas that is otherwise vented to the atmosphere might be marketed.

Emissions from rice paddies could be reduced by developing new varieties that produce high yields under rain-fed conditions. Methane emissions from ruminant animals could be controlled through better feeding strategies. One suggested option is to increase animal productivity through hormones or feed additives, but such practices pose both serious socioeconomic problems and possible human health risks. Eliminating subsidies that support large cattle populations would also reduce this methane source. Additionally, animal wastes can be tapped for biomass energy. Under controlled conditions, wastes can be digested, and methane can be extracted for use; many dairies and hog farms power themselves on their own manure.

Similarly, we can recover the methane gas emitted by landfills. Of the 5,000 landfills in the United States, about 1,500 vent methane but only about 120 collect methane for energy recovery. At New York City's Fresh Kills landfill on Staten Island, enough methane is recovered to provide energy to 12,000 Staten Island homes. Recovering methane from landfills and using it for energy should be the rule, not the occasional exception. One hurdle is state laws that subject resource recovery projects to unlimited liability for potential contamination problems at a landfill. These discourage methane recovery.

There is no reason for most organic waste (such as lawn

clippings and food scraps) to wind up in landfills, where it takes up precious space and produces methane as it decomposes. Local and state governments can establish programs to separate out this waste and compost it. Problems to be met include: what compost techniques to use (perhaps borrowing from widely used sludge composting techniques), and how to market the product. For a start, garden and park wastes could be prohibited from landfill disposal.

Other sources of methane are coal seams exposed during mining. States can require methane recovery at coal mines; Virginia has already enacted such legislation. Methane is also vented and flared during oil production and lost to the atmosphere when distributed through leaky pipes and valves. The amount of methane lost this way is unknown. But better recovery efforts and improvements in the distribution system would cut such emissions.

Nitrous Oxide

One way to curtail emission of nitrous oxide is to reduce substantially the use of nitrogen fertilizers, which are used extensively in American agriculture. Several management techniques can increase fertilizer efficiency, including adding nitrification inhibitors to fertilizers in order to decrease nitrogen loss; using slow-release or timed-release fertilizers to improve nitrogen uptake efficiency by plants; encapsulating fertilizer pellets to decrease water solubility and incorporate organic matter into the soil.

Educating farmers on such methods to minimize the use of nitrogenous fertilizer without sacrificing yield would also help cut down on nitrous oxide emissions.

Global Cooperation

While it is getting its own greenhouse in order, the U.S. government also needs to push for an international treaty limiting emissions of all the greenhouse gases discussed here. This is not an easy task, and not one that the current administration seems interested in pursuing. Still, recent international cooperation to curb the threat posed by chlorofluorocarbons is a heartening example that such accords are possible.

In contrast, while several western European countries have pledged to control their emissions of greenhouse gases, the United States remains the odd man out. During the deliberations of the Intergovernmental Panel on Climate Change in 1990, the United States found itself allied with Middle Eastern oil-producing countries and the Soviet Union as the leaders of a bloc fighting to downplay the threat of global warming. We need to do better.

One area in which this country can lead the world is technology transfers to the Third World. Whatever the industrialized West manages to accomplish on the greenhouse gas front can be wiped out by growth in less developed countries. The West now emits the lion's share of greenhouse gases, on both an absolute and a per capita basis. But the developing nations are not far behind (after the United States and the Soviet Union, the three leading emitters of carbon dioxide in 1987 were China, India, and Brazil), and their combination of population and economic growth threatens to propel them past the industrial nations as greenhouse threats. China alone is planning to build, over the next 10 to 15 years, 100 million refrigerators and to install 100 gigawatts of coal-fired electric power plants.

This threat is what José Lutzenberger calls "an elephant in the basement." In 1950 the developing world contributed 7 percent to global carbon dioxide output; by 1987 their share had risen to 28 percent. By 2020, developing nations could account for more than half of total carbon emissions.

Attempts to control worldwide emissions have been met with reactions ranging from recalcitrance to outright anger on the part of less developed nations. They feel that they are now being asked to pay the price for the West's profligacy. "Less developed countries require energy to escape poverty," says Lutzenberger.

To bring the developing countries on board will require assurances that they can achieve their economic goals in an environmentally benign way. That will require "green" technology—everything from small solar thermal generators capable of electrifying a single town in India to compact solar refrigerators suitable for storage of medicine and vaccines, food, and ice. These technologies exist, but at prices of, for instance, $3,300 to $8,500 for one solar refrigerator like those already installed at sites in the Marshall Islands and Thailand, the affluent West will have to help put such devices within reach of developing nations. It would be in our own best interests to give poorer nations, for free, technologies to eliminate chlorofluorocarbons or replace coal-fired utilities with those running on renewable resources. This approach was embodied in the renegotiated 1990 Montreal Protocol—the international agreement to phase out certain CFCs and halons. A fund generated by donations from wealthy countries will distribute $140 million over a three-year period to developing countries. These funds are supposed to assist them in ceasing their use of ozone-depleting chemicals.

Another approach would be to set up an international greenhouse gas–emissions trading system. The first task would be to set limits on greenhouse gas emissions. Subsequently, each country could use any means necessary to meet such limits. If a country achieves lower emissions than the regulated limit, it could then acquire income by selling its excess "credits" to another country that is unable to meet its prescribed emission limit. This system would provide incentives for energy efficiency and the efficient management of greenhouse gases. It would also provide a means for technology transfer. Industrialized countries could receive emission credits as partial payment for the installation of more efficient technologies. Also, a developing country with large forest resources and a burgeoning industrial sector could earn substantial emission credits by permanently preserving its forests. It could sell the credits to fund the renewable resource energy projects necessary to meet its growing energy needs.

3 | What You Can Do to Beat the Heat: 29 Steps

FEW OF US have much clout when it comes to, for instance, changing the lending practices of the World Bank in order to foster greenhouse-benign development in the Third World. But that is not to say that individuals cannot have any impact on global warming. Far from it. We can have an enormous impact. Individuals are directly responsible for about 40 percent of the energy used in the United States, for such things as heating and cooling our homes, powering our cars, running our appliances, and lighting our rooms. The other 60 percent of energy use is harder to affect, but not impossible. That energy goes to such uses as manufacturing, mass transit, and commercial use, so by our purchasing decisions and our behavior at work we can affect a lot of this indirect use of energy.

While government, utilities, and industries start acting to mitigate the threat of global warming, each of us can start now to reduce our own contribution to the greenhouse effect. Don't fret that you'll be asked to freeze in the dark and drink warm beer. Most of the steps outlined here call for some awareness and a modicum of effort—such as keeping your car tires properly inflated or finding a more efficient light bulb—but no sacrifices and no radical changes in the way we live, work, and play. Just the opposite. Many of the strategies that can help control the emission of greenhouse gases will mean getting the same service, such as refrigeration or lighting, at lower energy cost. This, of course, can translate into dollar savings, too, which can be put to better uses than underwriting the dividends paid by utility companies. Amory Lovins has noted that such energy-saving schemes are not a free lunch, but are "a lunch you're *paid* to eat."

Even a bottom line–oriented utility such as California's Pacific Gas and Electric Company now accepts the wisdom and economics of saving energy. Prodded by the EDF, which developed a computer model to demonstrate the relationships between conservation and electricity costs, PG&E now plans to meet fully three-quarters of its new electricity needs in the coming decade with efficiency—by getting both residential and business customers to consume less electricity without cutting back on equipment or burning fewer lights. Why? Richard Clarke, chairman of PG&E, says it is a simple business decision: "Energy efficiency costs three or four cents a kilowatt-hour compared to at least twice that for a new power plant." PG&E now rents a computer model from EDF for use in its regular planning.

Even if scientists are dead wrong about global warming, there are many good reasons for individuals to reduce their share of greenhouse gases. Some of the steps outlined here would reduce the problems of solid waste and the attendant increases in local taxes, acid rain, depletion of stratospheric ozone, and exploitation of precious natural resources for coal, oil, and gas. The 29 described steps below are simple, cost-effective ways to use energy more efficiently. The appendix lists specific sources for further information on the topics covered in this chapter.

Your first step is to figure out exactly how much carbon dioxide you are directly responsible for. Below is a worksheet to help you estimate your direct contribution to the greenhouse effect due to emissions of carbon dioxide.

How to Calculate One's Own Contribution of Carbon Dioxide to the Atmosphere

The average American releases about 20 tons of carbon dioxide into the atmosphere each year. The figure is very rough—it includes transportation and home energy use, but doesn't include industrial and government production. Use this table to make an equally rough estimate of your and your family's production of carbon dioxide. Where can you begin to cut back?

1. Calculate the number of gallons of gasoline you used last year (number of gallons = number of miles traveled ÷ miles per gallon for your car): _____
Multiply by 20 (burning one gallon produces 20 pounds of carbon dioxide): _____

2. Number of miles flown last year: _____
Multiply by 0.5 (flying one mile produces about one-half pound of carbon dioxide per passenger): _____

3. Now dig out your past 12 months of utility bills. Add up the number of kilowatt-hours of electricity used at home last year: _____
Multiply by: 2 if coal generated (1 coal-generated kwh produces 2 pounds carbon dioxide)
0 if produced by hydropower or nuclear power.
1.25 if by natural gas.
1.7 if by oil
(Check with your electric utility to find out how your electricity is generated. It may vary over time.)

_____

4. Number of cubic feet of natural gas burned (in hundreds): _____ ccf (hundred cubic feet)
Multiply by 12 (1 ccf produces 12 pounds of carbon dioxide; 1ccf = 1 therm = 100,000 btus): _____

5. Number of gallons of fuel oil burned last year: _____
Multiply by 20 (1 gallon produces 20 pounds carbon dioxide): _____

SUBTOTAL: _____

Now, double the subtotal to take account of carbon dioxide released in the production of food, goods, and services each person purchased.

GRAND TOTAL: _____

Now that you know your approximate carbon dioxide contribution, the next step is to try to pare it down.

On the Road

1. Whenever possible, walk, bike, carpool, or use mass transit. Cars and light trucks produce 20 percent of all carbon dioxide emissions in the United States. The average American car emits 7.5 tons of carbon dioxide per year (based on 15,000 miles). An inescapable fact: a car or truck that is not running produces no carbon dioxide. Just by carpooling regularly with one other person, you can cut your transportation-related carbon dioxide emissions by up to half, and save money, too. If there were just a 10 percent increase in the number of commuters who carpooled, we would save 7.5 million gallons of gasoline, and 15 million pounds of carbon dioxide emissions every day. (In fact, there is so much extra room in our cars that we could carry everyone in western Europe around with us and still have room for luggage.) The savings would be even greater if we added in the gasoline saved through the reduced congestion. Cars burn an estimated 3 billion gallons of gasoline a year—and emit 60 billion tons of carbon dioxide as they idle in traffic.

2. Drive smart. Stay within the speed limit; speeding can increase your annual gasoline bill by 20 percent and add that same amount more of carbon dioxide to the atmosphere. Texaco, which in 1990 launched a nationwide energy-efficiency campaign called "Let's Put Our Energy into Saving It," estimates that if all Americans drove no faster than the legal limit, the country would save 4.2 million gallons of gasoline every day. That means a savings of 84 million pounds of carbon dioxide daily.

- Avoid jackrabbit starts; they can drop your fuel efficiency as much as 2 miles per gallon.
- Use the air conditioner sparingly; that too can save up to 2 miles per gallon.
- Radial tires can cut your gasoline bill, and your contributions of carbon dioxide, by 10 percent. Keep your tires inflated to the manufacturer's specifications. If your tires are underinflated, you will lose up to one mile per gallon of fuel economy. If every car had firm tires, we would save another 4.2 million gallons of gasoline a day (84 million pounds of carbon dioxide), estimates Texaco.
- A regular tune-up can save 8 percent in gasoline costs and the same amount in carbon dioxide.
- Don't idle unless absolutely necessary. Keeping the engine running for longer than one minute wastes up to one gallon of gasoline per hour. You'll use less gasoline if you turn off the engine and restart it when you're ready to go again.
- Keep the air filter clean. A clogged one can cut fuel economy by another 1 mpg.
- Don't drive around with everything but the kitchen sink. Take what you don't need out of the trunk. Every extra 100 pounds in the car consumes up to 0.2 miles per gallon.

3. Make sure your automobile's air conditioner is in good shape. Fix leaks, don't just recharge the air conditioner; automobile air conditioners put more chlorofluorocarbons into the atmosphere than any other single source. The major routes are steady, slow leakage and rapid venting when the air conditioner is serviced. When servicing is needed, have the refrigerant recycled, rather than dumped into the atmosphere. A few service stations, dealerships, and air-conditioning specialty shops may have the necessary CFC-recycling equipment. Several models have been certified by Underwriters Laboratories; others are currently being tested. General Motors has pledged to make recovery and recycling equipment available to its dealerships in 1991. Nissan has said it will eliminate the use of CFCs in its air conditioners by 1993. Meanwhile, shop around to try to find a service station that will recycle the refrigerant from your car's air conditioner when it's repaired.

4. When it comes time to buy a car, choose one with high fuel efficiency. The increasing fuel efficiency of automobiles from the early 70s to the mid-80s has held gasoline consumption steady over the past 15 years despite an increase in the number of miles traveled by cars. A car that gets 20 miles per gallon (current U.S. average) will emit approximately 50 tons of carbon dioxide over its lifetime; a 28 mpg car (new car average) will emit 35 tons, and a 40 mpg car will emit only 25 tons. Over the average lifetime of an American car (100,000 miles), a 40 mpg car will also save approximately 2,500 gallons of fuel compared to a 20 mpg car.

Every year the U.S. Environmental Protection Agency and the U.S. Department of Energy produces a "Gas Mileage Guide" (see Appendix) that lists the fuel economies of all new cars. Check this before shopping for a new car; all new-car dealers are required by law to have a copy. Each April issue of *Consumer Reports* is devoted to automobiles and has fuel economy information for most new cars.

In the Home

One-fifth of all the energy used in the United States is used at home. In 1984 space heating accounted for about 40 percent of total U.S. residential fuel use, making it a good target for efficiency measures. The following pie chart illustrates where all home energy goes.

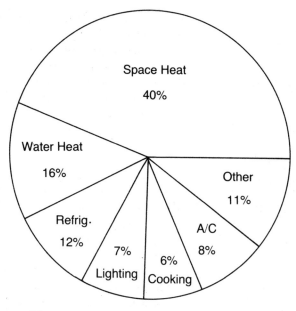

FIGURE 6. *Home energy use*

Space Heating and Cooling

Making your home comfortable accounts for half your energy bill, but there are some simple steps you can take to reduce the expense. The goal is to keep the warm or cool air where it belongs—in the house instead of leaking out windows, doors, walls, and ceilings. One-third of the heated air in a home is lost through windows. To heat that lost air requires the energy equivalent of nearly the annual output of the Alaskan pipeline, or about 5 percent of *all* the energy used annually in this country, which amounts to about 179 million tons of carbon dioxide that could be avoided.

5. **Weatherize your home or apartment.** For a very small investment, you can begin to cut your heating and cooling expenses and reduce the burning of fossil fuels, if that's where the

energy comes from. Pulling the shades down over your windows at night can save several dollars per window per year. Drapes work even better, especially with a valance on top to trap air. Next, weatherstrip windows and doors. Various methods are available, all of which help make the window or door more airtight. Rope caulk is inexpensive and easy to apply, but is meant for windows that needn't be opened. Several kinds of plastic or metal strips are available to fit around doors and windows that must open. Don't forget the bottoms of doors when you are weatherstripping. Permanent door sweeps are available, including some high-tech models that retract when the door is open. A simpler alternative is to place a draftguard—a weighted, closed cloth tube—against the bottom of the door. If you heat with electricity, caulking and weatherstripping can reduce energy enough to keep 1,100 pounds a year of carbon dioxide from being emitted by the power plant. In general, the rule is to caulk joints that don't open, such as where a wall meets the outside edge of a window frame, and to weatherstrip places with moving parts, such as where windows and doors close in their frames.

6. Lower your thermostat settings. Pushing it from 72 degrees to 68 degrees can save about $80 on your annual heating bill, and even smaller setbacks can help reduce greenhouse gas emissions. Although fuel and carbon dioxide savings depend on how cold it is outside and what fuel you heat with, as a rule of thumb every 2 degrees you lower your thermostat in the wintertime will cut carbon dioxide emissions related to heating by 6 percent. The savings can be painless if they're made at night (invest in a good comforter or quilt, but not an electric blanket!).

7. Modernize your windows. Windows are usually the weak points in a home's insulation; they deserve special attention in any home energy saving plan. Your strategy will depend on where you live and the state of your present windows. In a warm climate, single-glazed (single pane) windows may be sufficient, especially if they are properly shaded from summer sun and weatherstripped. In colder climates, windows should be at least double-glazed. How to know if improvements are necessary? The windows are probably in adequate shape if they don't have much condensation on their inside surfaces on cold days. The better an insulator a window is, the warmer its inside surface will be on a cold day. Another test, of course, is comfort. If there are drafts, the windows need attention.

If you're building a new home or have to replace windows that are in bad shape or inadequate for the climate, install energy-efficient windows. Some on the market now retain heat three or more times as well as single-glazed windows. The insulating ability of windows is often rated according to the same scale as wall and ceiling insulation. A typical wall may be an R-11 insulator; a single-glazed window is R-1, meaning that a wall across the same space would be 11 times better at resisting heat flow. Double-glazed windows are roughly R-2. Modern glazings include more than just additional panes of glass. Some windows ("low-emissivity" or low-E) are coated with a film that reflects infrared radiation (heat); in cold climates, sunlight warms the interior of a room, and the low-E windows reflect radiating heat back into the room. Low-E windows had saved U.S. consumers $14 million a year by 1985. By 1995 low-E windows are expected to save $120 million in annual heating costs. Other windows with multiple panes may have argon gas in place of air in the space between the panes. Argon conducts heat more poorly than air.

Many companies offer R-3 and R-4 windows that are only moderately more expensive (10 to 20 percent) than double-glazed windows. Installation costs are the same in any case, roughly equal to the cost of the windows. When comparing window costs, try to factor in the yearly energy savings possible with the more efficient windows. The payback period for the better windows may be relatively short, about three to five years. The windows, meanwhile, should last about 20 years, saving money every year. Don't forget to examine the frame when purchasing windows. Aluminum frames conduct heat well and will nullify much of the savings gained by the low-E windows. Look for wood or vinyl frames instead (or for aluminum frames with a *thermal break*—a nonconducting layer—in them).

Once you have weatherized your windows, the next step is to add storm windows. Air trapped between the primary window and the storm window acts as insulation. If either window leaks air, the insulating ability of the storm window will be largely lost.

8. Use a fan instead of an air conditioner, if possible. Fans use one-tenth the energy of air-conditioning. If you need an air conditioner, search for an energy-efficient one. All air conditioners are sold with a label providing the energy efficiency rating in terms of EER (for room air conditioners) or SEER (for central air conditioners); the higher the EER or SEER, the better. Use the American Council for an Energy-Efficient Economy's *The Most Energy-Efficient Appliances*, an annual guide that lists the top makes and their efficiencies (see Appendix).

Since about 64 percent of our homes have air conditioning, and every year about 8 million new units are sold, we can give the greenhouse a break by at least using them wisely.

- Logically, the unit cools best if it itself is cool: keep it in the shade if possible, and you'll cut its electricity consumption by about 5 percent.
- If you've got whole-house air-conditioning, close the doors and vents to the rooms you don't have to cool.
- Don't set your air conditioner's thermostat to freezing levels; the recommended summertime setting is 78 degrees. With central air, that reduces your electric bill 8 percent compared to a setting of 75 degrees Fahrenheit, and cuts carbon dioxide emissions associated with the generation of that electricity by the same amount.
- Check every spring to make sure the air conditioner's coils are clean and straight; that will make the unit work more

efficiently and thus save energy. Similarly, keep the filter clean; a clogged filter will consume 5 percent more energy than the unit really needs. If you've got disposable filters, toss them every month or so; permanent ones should be cleaned according to the manufacturer's directions.

9. Make sure your home has adequate insulation. Many older homes do not have enough, especially in the attic. You can measure the insulation yourself or have it done as part of an energy audit (references listed in the Appendix help translate the amount of insulation to insulating ability). Adding insulation may have a payback time as short as a few years. You can do some jobs yourself. One good place to start is with the heating ducts, which almost always leak (warmed) air as well as heat if their thin metal walls are not insulated. So do insulate them, preferably with two-inch-thick fiberglass, and tighten them up. That alone can save up to 10 percent of your heating costs. Clean around the joints and seal them with duct tape.

The do-it-yourselfer can also tackle basement and attic insulation. Walls are best left to a contractor: insulating them may require drilling between the studs and blowing in insulation.

10. If you find yourself building a new home in the frostbelt, explore the possibility of *superinsulating* it. That means adding more than the recommended amount of insulation (sometimes doubling the typical R values for walls and attic), using higher quality, weatherized windows and doors, and making other heating-efficiency improvements over typical homes. Depending on the experience of the builder, a superinsulated home may cost up to 5 percent more than a normal home. The payback period depends on climate and heating and cooling costs, but it can be just a few years, because space conditioning bills sometimes drop to nearly zero. (Some of the best insulators are rigid foam products that currently contain CFC-11, a gas that both destroys the ozone layer and contributes to global warming. Check with an insulation contractor for alternative insulation products that may be used in greater quantities to achieve an equivalent R-value, or ask if the foam manufacturer uses an alternative chemical.)

The increasing popularity of superinsulated construction has helped focus attention on *sick building syndrome*, wherein reduced ventilation combines with sources of indoor air pollution to cause health problems for a building's inhabitants. Superinsulation does not *cause* sick building syndrome. Rather, it can amplify existing indoor air quality problems, according to David Grimsrud of the University of Minnesota Cold Climate Building Research Center. He recommends attacking the problem by eliminating or reducing *sources* of pollutants. If you plan to tighten up your home, test for pollutants first. Radon is the most well known of the indoor air pollutants and is relatively easy to test for. Try to minimize the sources of pollutants (pressed-wood products, some foam products, some backings or glues for carpets and drapes, unvented gas and wood combustion, household cleaners, and humidifiers and air conditioners) in new construction. Tightening a home below ground can have the added advantage of helping to reduce the infiltration of radon gas from the soil into the basement.

If you can't eliminate all indoor air quality problems by controlling sources, install a heat-recovery ventilator—an air exchange device that uses heat from exhaust air to warm incoming fresh air, and thus minimizes ventilation-related heat loss.

11. Get a home energy audit. Many utilities offer them for free or at low cost. An inspector will point out areas in your home where you can save energy and money. Some utilities or state agencies may help finance the purchase of energy-efficient appliances.

When you use hot water inefficiently, you might as well be pouring money down the drain—and pumping carbon dioxide into the air. Hot water is one of the easiest things to abuse and one of the easiest to conserve—and you don't have to take cold showers.

12. Turn down the temperature of your hot water heater to about 120 degrees Fahrenheit (or the "low" setting). Each 10 degrees Fahrenheit reduction can cut water heating costs 3 to 5 percent. Besides, how often do you run a faucet or shower without mixing cold water into the hot? (If your dishwasher doesn't have its own heating element, keep the water heater set at 140 degrees F.) If you heat water with natural gas, you will save about 40 therms a year and 440 pounds of carbon dioxide, calculates the National Audubon Society. For an electric heater, the savings are around 400 kilowatt-hours (check your utility bill to see how much of a savings in dollars this yields) and 600 pounds of carbon dioxide a year (more or less, depending on the mix of fuel used by your utility).

13. Insulate your water heater. Kits are available in home improvement and hardware stores for about $20; or you can use fiberglass insulation. Take care not to cover vents or controls. This will save you 20 therms and 220 pounds of carbon dioxide annually for a gas heater, or 700 kilowatt-hours and 1,100 pounds of carbon dioxide a year if you heat your water electrically.

14. Buy an efficient water heater. When you need a new one, there are a variety of models on the market. Their purchase prices can vary by a factor of 10 and the yearly energy cost by a factor of three, so it pays to shop around and compute life cycle costs. (*Saving Energy and Money with Home Appliances,* listed in the Appendix, shows how to compute these costs and gives helpful information.) If you buy a water heater with a storage tank, make sure the tank is well insulated. "Demand-type" water heaters don't have storage tanks but heat water only when you need it, at an energy savings of 20 to 30 percent.

15. Install low-flow showerheads on all your showers. They're available from some plumbing supply stores or in catalogues (see Appendix). The better ones can reduce water usage by more than one-half. They work by mixing air with the water to maintain a vigorous stream. (Avoid simple flow restricters, which just cut water volume and yield for a less-than-satisfying shower.) Similar aeration devices fit sink faucets. Low-flow fixtures can reduce total hot-water use by 10 to 15 percent and the annual utility bill by $50 for an electric water heater or $20 for a gas water heater. This will save you 8 therms and 80 pounds of carbon dioxide a year if your water heater operates on gas; the savings are 200 kilowatt-hours and 300 pounds of carbon dioxide with electric water heaters.

Saving water of any temperature can help curtail greenhouse gas emissions because energy is needed to transport, filter, and otherwise treat the water that finds its way to your tap. California's PG&E calculates that a low-flow aerator each on one kitchen and one bathroom sink faucet will save a typical family of four up to 280 gallons per month—3,300 gallons a year.

16. Wash clothes in warm or cold—not hot—water. As much as 90 percent of the energy used in washing clothes in a machine goes to heating the wash water, according to the Association of Home Appliance Manufacturers. So use the coolest water possible. Switching from hot to warm cuts energy consumption about 50 percent per cycle. Always rinse in cold water, since rinse-water temperature doesn't affect cleaning.

If your machine has a water-level selector, use it. Filling the basin with more water than necessary is wasteful. If yours does not, then wash only full loads. Washing machines account for about 15 percent of the water used in homes that have them—between 32 and 59 gallons every wash cycle, says PG&E. Another tip: use the "delicate" setting if your machine has it; that doesn't work the motor as hard and consumes less electricity.

17. Buy energy-efficient appliances if at all possible. Several major appliances (refrigerators, freezers, water heaters, clothes washers, dishwashers, and room air conditioners) sport Energy Guide labels, required by federal law. The labels give estimates of yearly energy cost or can help you compute your own estimate. The most efficient appliances often have a higher purchase price, but can save money over the life of the product. Remember that *the refrigerator is the appliance that consumes the most* *electricity in the home,* so when you have the opportunity (especially if you are building a home), purchase the most energy-efficient model possible. As an example of the improvements that have been made lately, new refrigerators and freezers are, on average, about 75 percent more efficient than those produced in the early 70s. The 125 million refrigerators and freezers in U.S. households require 30 1,000-megawatt power plants to run. If all those refrigerators and freezers were 1975 models, they would require 50 such power plants. On the other hand, if they were all replaced with the most efficient models available, we could shut down all but 18 of those power plants. Replacing the typical refrigerator in use today with the most efficient one on the market (assuming no change in size or features) would cut the average household's annual electric bill by $21.

In general, models with the freezer on top consume less energy than do side-by-side refrigerator/freezers.

18. Run the refrigerator you have as efficiently as possible, if for no other reason than that it accounts for one-quarter of your electricity use if you're a typical city dweller. There are several tricks to minimizing the unit's energy use:

- Keep the condenser coils behind or under the refrigerator clean. When dust builds up on them, they can't eliminate heat to the air as efficiently, so the compressor motor has to work longer, drawing more electricity.
- Keep the thermostat at between 38 and 42 degrees Fahrenheit in the refrigerator, and between 0 to 5 degrees Fahrenheit in the freezer. Temperatures a mere 10 degrees colder than these can jack up the refrigerator's energy consumption by 25 percent.
- Check to see if coolness is leaking out (or warmth is leaking in). Close the door on a sheet of paper and, if the paper slips out easily, adjust or replace the gasket (rubber seal) around the door.
- If your refrigerator has a power-saver switch, usually on an inside wall, use it. This controls tiny heaters built into the walls that keep water droplets from condensing in hot,

humid weather. Except on the muggiest summer days, turn on the switch (and keep it on then, too, if you don't mind some "sweat" running down the refrigerator walls).

- Stock up. Although air needs to circulate in order to keep food cool, keep both refrigerator and freezer relatively full. The reason—food retains the cold better than air does. Think of those blocks of butter and packages of frozen spinach as ice cubes helping to keep the temperature down. Similarly, if you have to defrost a turkey, for instance, leave it in the refrigerator for a few days. That's safer from the point of view of food poisoning and energy efficient as well. The turkey cools its surroundings as it thaws.
- Let warm foods, like cooked pudding or tea that needs chilling, cool down a little before you put them in the refrigerator.

19. Unplug the second refrigerator or freezer—if it's largely empty—in your garage or basement. You'll save carbon dioxide and money too! For example, a 12-year-old model that uses 1,500 kwh per year contributes over a ton of carbon dioxide. Plus, you'll save $50 to $100 a year even if the refrigerator is an efficient one.

20. Don't overdry clothing in a clothes dryer. Look for models with moisture-sensor control, a cool-down cycle, and (if gas powered) electronic ignition instead of a pilot light. Air dry your clothes whenever possible; put up a clothesline and use solar power. Keep the dryer's lint filter and exhaust hose clean and unobstructed. If air circulation is blocked, the machine consumes more energy. As with washing, dry full loads, but don't pack the machine. If air cannot circulate around the clothes freely, they take longer to dry.

21. Run dishwashers only when fully loaded. Use an air-dry cycle if available (or crack open the door when the wash cycle is over). Depending on how you wash dishes, an efficient dishwasher can use *less* energy (because it uses less hot water) than hand washing. However, if you fill the wash basin first and don't let the water run, you'll use less water than a dishwasher. A worthwhile feature to look for in a new dishwasher is a *booster heater*, which heats incoming water to 140 degrees F. for best washing results. Then you don't need to keep your water heater set at that temperature.

22. Buy a gas range without a pilot light. Electronic ignition can cut the stove's gas use by 40 to 50 percent and save $25 a year.

Lighting

The standard incandescent bulbs (known as lamps in the trade) that you use to light your home may seem like a bargain, but they're not. The typical A-line lamp (standard incandescent bulb) costs 75 cents or so and lasts 750 to 1,000 hours. During its short lifetime, it accounts for over 100 pounds of carbon dioxide and consumes 60 to 100 kwh of electricity costing $5 to $6. See the table below to calculate the real cost of lighting.

Calculating the Real Cost of Lighting

Use this worksheet to calculate the cost per thousand hours of using any light bulb you are considering buying.

(1) What price do you pay for electricity per kilowatt-hour? $ _____/kwh

(2) How many watts are used by the light bulb?   _____/watts

(3) Multiply (1) by (2):
    $_____kwh × _____watts = $ _____

(4) What is the rated lifetime of the bulb?   _____ hrs

(5) Divide this by 1,000: _____ ÷ 1,000 =   _____

(6) What did you pay for the bulb? $ _____

(7) Divide (6) by (5):
$ _____ ÷ _____ = $ _____

(8) Add (3), operating cost, to (7), the
purchase price: _____ + _____ = $ _____ 1,000 hrs

Total cost per 1,000 hours $ _____

If labor costs for bulb replacement are significant, as they can be for apartment building hallways, for example, they should be added in at step (6).

Incandescent bulbs make light very inefficiently—electric current flows through the filament, heating it up until it glows. Only about 10 percent of the energy a light bulb uses makes light; the rest is lost as heat.

Fluorescent lights are much more energy-efficient than incandescent lights. They put out the same amount of light at roughly a quarter of the wattage of incandescents. New *compact fluorescent* lamps screw directly into a normal socket. They last about 10 times as long as a standard incandescent and consume about one-quarter of the power. These new fluorescent lamps give off a more pleasant light than the old kind, with little flicker or hum. The disadvantages are the high initial cost, that their size and shape make it difficult to fit them into closed lamp fixtures, and that they are not yet as easily found as incandescent lamps. However, they are certainly worth looking for.

It's not that compact fluorescent lamps are not readily available. They are, but they aren't yet typically stocked in the light bulb section of the supermarket. It was only in 1990 that a major national store—Sears—began stocking compact fluorescents. Check the Light Bulbs and Tubes heading in the *Yellow Pages* to find specialty lighting stores that sell compact fluorescents. These lamps are expensive—from $10 to $30 each—but in fixtures that receive a lot of use, they can save money in energy costs because they can last 10 times as long as standard incandescents. For example, compared to the standard 75-watt bulb that costs $6.55 per thousand hours (which includes the cost of electricity), a 20-watt compact fluorescent costs $4.08, a savings of $2.47. So, lighting a room for 10,000 hours with standard incandescent lamps would cost $25 more than using a compact fluorescent, not including labor costs. Also, remember that in addition to the money saved, approximately a half-ton of carbon dioxide is saved by using the 20-watt compact fluorescent instead of several incandescents over the same 10,000 hours.

In fixtures where compact fluorescent lamps aren't practical, there are still alternatives to standard A-line incandescents. *Tungsten-halogen* lamps are now available that fit regular incandescent sockets. These lamps use a third less electricity than a standard incandescent bulb, last about four times as long, and cost about $2.50 to $3.00 each. *Krypton-filled* incandescents, often called supersavers or watt misers, deliver the same light as a standard incandescent with a 5 to 10 percent energy savings.

23. Replace the three most-used bulbs in your home with compact fluorescents, and the remaining bulbs with tungsten-halogen lamps. You will cut your annual electric bill by about $30, and will need to change these bulbs a lot less often. If every American home took this step, it would reduce this country's total carbon emissions by about 1 percent.

Be cautious about claims for energy savings when buying bulbs. An unambiguous guide to a bulb's efficiency is its light output per watt of power consumed. Light output is measured in *lumens*, usually reported on the box along with the bulb's wattage. To compare two bulbs, divide lumens by watts for each one. Some long-lived incandescent bulbs, ironically enough, have lower efficiencies than standard A-line bulbs.

No matter what type of bulbs you have, keep them clean. According to PG&E, dust on a bulb or light fixture can cut the luminosity by 10 percent, making it seem as if you need a brighter, higher-wattage light. You can also save on wattage by being a smart interior decorator. White walls reflect 80 percent of the light that strike them; at the other extreme, black walls reflect a mere 10 percent. Also, despite the popular misconception, it does not take more energy to switch a light back on than to keep it burning for an extra few minutes. When you're not using an incandescent light, *turn it off*. Also, whenever your fixtures permit it (check their wattage ratings), use fewer high-wattage bulbs instead of many low-wattage ones. For example, one 100-watt bulb draws the same amount of electricity as four 25-watt bulbs, but emits twice as much light.

Recycling

Most of what gets tossed into municipal landfills isn't really garbage, but raw materials. Aluminum cans can be used to make more aluminum cans. Newspaper is easily turned into more newsprint or other paper products. Some plastics make good feedstock for the production of more plastics. With many landfills in the United States running out of room and with much of our trash being nonrenewable material, recycling makes common sense—even without global warming.

In practical terms, the U.S. Environmental Protection Agency predicts that about 50 million tons of newsprint, containerboard, and glass, plastic, and aluminum packaging will be landfilled or incinerated in the year 2000. If that material were recycled instead, the energy this saved would make the equivalent of 10 billion gallons of gasoline available to use in other ways. Put another way, it would be as if more than 5 million Americans stopped putting carbon dioxide into the air.

In the best case—that is, if recycled products are made back into what they were—the energy savings mount up quickly. By recycling a one-foot-tall pile of newspapers, you save enough energy to take seven hot showers. For every glass soda bottle you recycle, you've saved enough energy to run a television for

90 minutes. Discarding an aluminum can is as wasteful as filling it halfway with gasoline and spilling it down the sewer.

24. Buy goods that can be recycled or reused. Find out what materials are recyclable in your community and choose products packaged in those materials. (Most packaging materials are recyclable in principle. The key is what recycling options are available where you live.) Otherwise, choose products whose containers you can reuse or that have less packaging.

Personal Choices

25. Consider eating less meat. There aren't many ways in which individual actions can reduce methane and nitrous oxide emissions. One effective method, if practiced on a large scale, would be to modify our diets with the Earth in mind. Eating less red meat can reduce the size of cattle populations, which would reduce methane emission directly (fewer cow belches and flatulence) and indirectly (less manure). In addition, since cattle eat much of the corn grown in the United States, fewer cattle mean less corn. Reduced acreage for corn—a heavily fertilized crop—would thus reduce a source of nitrous oxide emissions. Substitute poultry for red meat, or grains and vegetables for either.

Tree Planting

Up until now, the steps have addressed ways to reduce the amount of greenhouse gases we add to the atmosphere. Another approach is to increase the rate at which greenhouse gases are removed from the atmosphere. One of the best ways to do that is to plant trees. Trees do more than consume carbon dioxide. In urban areas, by providing shade, evaporative cooling, and wind breaks, trees can reduce greenhouse gas emissions from heating and cooling by 15 times as much as they themselves absorb directly.

Tree planting is an economical way to save energy. Compare the expense of reducing energy demand by a variety of methods: by planting urban trees, it costs about one penny to save a kilowatt-hour of energy (which itself only costs about 8 cents). It takes about two cents to save that energy by improving appliance efficiency, and 10 cents to do it by improving the efficiency with which the energy is supplied.

26. If you live in an urban area or a warm climate, plant trees. People in cool climates should plant deciduous trees on the south and west sides of their homes, to provide shade in summer and allow solar heating in winter.

Which trees best counter greenhouse warming? No single species is best suited for all climates, soils, and other factors. Choose a tree that is native to your area and will provide a lot of shade. Because the net amount of carbon dioxide that a tree eliminates depends on the amount of weight it adds, you should choose a tree that grows fast. About half the dry weight of a tree is carbon. Contact your state or local forester or extension agent for recommendations (or see the Appendix for more information).

At Work

27. Check with your office building manager to make sure everything that can be done is being done to save energy. Look around your desk or office at work. Are there lots of lights burning for long periods of time when no one is in the area? Is the office kept too cool in the summer or too warm in the winter? Are there large expanses of single-glazed windows? The typical American office is very energy inefficient. Some simple steps can greatly increase the energy efficiency in almost any office or workspace. If you work for a large corporation, chances are it has someone on staff whose responsibility is plant management. Because of the economic incentive to save money, that person may already be aware of many ways to improve energy efficiency. A smaller company is less likely to have an energy expert on staff, but there are many consulting firms that can advise on employing energy-efficiency measures.

28. Replace incandescent bulbs at work with fluorescent fixtures wherever possible. Many lights in commercial establishments—in hallways and entryways, or illuminating the grounds or building—burn long hours, and improving their efficiency would have substantial benefits. As a result, lighting of commercial buildings accounts for at least 25 percent of the buildings' energy use, as against 10 percent in homes. Even all-fluorescent lighting can be made more efficient. For example, there are about 850 million standard inverted fluorescent fixtures (with four fluorescent tubes) in commercial buildings. These fixtures, known as troffers, use about 184 watts but can be easily converted to use half that by removing two bulbs and installing better reflectors and a more efficient ballast.

Because incandescent bulbs produce so much heat, offices with lots of them almost always find that their air-conditioning energy use goes down upon a switch to fluorescent lighting.

Improvements in the color quality of newer fluorescent lights mean that your office doesn't have to have a washed-out look to be energy efficient. New solid-state high-frequency ballasts allow fluorescents to be dimmed—even automatically in response to increasing daylight. If the people in your office don't remember to turn lights off when they leave a room, there are sensors that can turn lights on automatically in response to movement and off with a sound (or lack thereof).

If your building hasn't had an energy audit, we suggest that one be conducted. If it has had one, find out whether the recommended measures have been followed.

Vote Green

29. Write or call your local, state, and federal representatives. Tell them what you think and why. Considering the United

States' role in contributing greenhouse gases, actions taken by our government can have significant global effects. Despite what you may think, government officials do pay attention to what you tell them. Be informed about relevant legislative bills that are up for debate or policies that are set by your local or state governments, and voice your opinion. Also, pay attention to issues such as transportation, recycling, and efficiency standards. Your civic participation is critically important! Unless citizens make their voices heard loud and clear, representatives have little incentive to take *significant* action on legislative measures that have the potential to minimize global warming. Urge others who share your views to do the same. The National Environmental Scorecard published by the League of Conservation Voters is an important resource to learn about the voting record of your elected officials on environmental issues. Get a copy of it and use it during the next election.

Appendix: Publications and Information Services

Publications

The Most Energy-Efficient Appliances, published by the American Council for an Energy-Efficient Economy (ACEEE), 1001 Connecticut Ave. N.W., Suite 535, Washington, DC 20036.

This annual guide lists top-rated appliances by brand, model number, and energy usage or rating. Includes a worksheet to help compute life-cycle costs. An indispensable tool when shopping for a major appliance. $3.

Saving Energy and Money with Home Appliances, published by Massachusetts Audubon Society and ACEEE. Massachusetts Audubon is at South Great Rd., Lincoln, MA 01773.

Explains the options and energy-saving features of many appliances; covers how to choose, install, use, and maintain these appliances. Gives background information on energy usage and comparison shopping. This booklet is part of a low-cost energy series published by Massachusetts Audubon. Other titles include: *How to Weatherize Your Home or Apartment, Super-*

insulation, Financing Home Energy Improvements, Solar Ideas for Your Home or Apartment, All About Insulation, among others. The booklets are all under $5 and are well written and illustrated.

Consumer Guide to Home Energy Savings (1990) by Alex Wilson. Published by ACEEE in cooperation with *Home Energy* magazine, 2124 Kittredge St., No. 95, Berkeley, CA 94704.

This comprehensive guide supplies you with information on energy-saving products and shows you how to use them most effectively. It covers all aspects of home energy use and provides practical answers to commonly asked questions. (Includes reprinted information from the Massachusetts Audubon Society booklets.) $6.95. *Home Energy* is a bimonthly publication available for a yearly subscription of $45.

30 Simple Energy Things You Can Do to Save the Earth (1990), $5 postpaid from EarthWorks Press, 1400 Shattuck Ave., No. 25, Berkeley, CA 94709.

Brought to you by the same people who published *50 Simple Things You Can Do . . . ,* this useful booklet lays out helpful facts and suggestions for improving energy use at home.

Consumer Guide to Efficient Central Climate Control Systems, published by Air-Conditioning & Refrigeration Institute (ARI), 1501 Wilson Blvd., 6th Floor, Arlington, VA 22209.

This 32-page pamphlet explains what heat pumps and air conditioners are, how to get the most out of them, how to purchase these systems, how to add climate control to an existing house, and what efficiency ratings mean. ARI publishes several other free pamphlets about space conditioning that include information on energy efficiency.

Resource-Efficient Housing Guide (1987) published by the Rocky Mountain Institute, 1739 Snowmass Creek Rd., Snowmass, CO 81654.

An annotated bibliography of books and periodicals dealing with resource-efficient housing. Very comprehensive and includes a directory of helpful organizations. $15.

The National Environmental Scorecard, published by the League of Conservation Voters, 2000 L St. N.W., Suite 804, Washington, DC 20036.

The League is a national, nonpartisan political committee formed to help elect conservation-minded candidates to office. The *Scorecard* is published every two years. It contains the League's rating of all senators and representatives according to their votes on key environmental issues.

Gas Mileage Guide, available at no charge from the Consumer Information Center, Pueblo, CO 81009.

This guide is published annually in a joint venture between the Environmental Protection Agency and the Department of Energy. All new-car dealers are required to have a copy. Most public libraries should have a copy. The booklet lists the gas mileage of all new cars for which data is available by the publication deadline. A spring supplement includes cars the fall guide missed.

Global ReLeaf Action Guide, published by the American Forestry Association, P.O. Box 2000, Washington, DC 20013. Cost is $1.50 or free if you send a contribution of $10 or more. AFA will also send you tree-planting information if you make a request to the above address. For a $5 phone call (1-900-420-4545), AFA will send you the same information and plant a tree. The *Action Guide* informs the reader about the environmental importance of trees and forests, lists state coordinators of Global ReLeaf, and instructs how to plant a tree.

Carbon Dioxide Diet for a Greenhouse Planet: A Citizen's Guide for Slowing Global Warming (1990), published by the

National Audubon Society, 950 Third Avenue, New York, NY 10022. $4.95.

This booklet describes in detail how to calculate an individual's personal emissions of the major greenhouse gases.

Atmosphere, a publication of Friends of the Earth—International on Climate Protection. This quarterly newsletter covers national and international efforts to stop ozone depletion and slow down global warming. Available for a $15/year subscription fee from Friends of the Earth—United States, 218 D Street, S.E., Washington, DC 20003.

Information Services

Conservation and Renewable Energy Inquiry and Referral Service. Call 800-523-2929 (including Virgin Islands and Puerto Rico), 800-233-3071 in Alaska and Hawaii. Mailing address: Renewable Energy Information, P.O. Box 8900, Silver Spring, MD 20907.

A service of the federal Department of Energy, CAREIRS answers questions about conservation and renewable energy. Will refer those with detailed questions to other organizations or publications. Services available to anyone, without charge.

National Appropriate Technology Assistance Service. Call 800-428-2525, or 428-1718 in Montana. Mailing address: P.O. Box 2525, Butte, MT 59402-2525.

NATAS, also a service of the DOE, offers two kinds of help to people who are undertaking personal or commercial projects involving energy conservation or renewable energy: technical engineering information, including design, problem solving, cost benefit analysis, and locating local technical support for a project; and business assistance, including industry overviews and market trend analyses. All services are free, but certain types of questions can only be responded to within a two- or three-week period.

Catalogues

Seventh Generation. Write to: Seventh Generation, 10 Farrell St., South Burlington, VT 05403, or call 802-862-2999.

Mail-order products "for a cleaner, healthier environment." Catalogue includes some energy-saving devices.

Real Goods. Write to: Real Goods Trading Corporation, 966 Mazzoni St., Ukiah, CA 95482, or call 800-762-7325. This catalogue offers everything from batteries to wind generators and is available three times a year for a $25 subscription fee. The subscription includes the *Alternative Energy Sourcebook,* a $14 value. This 400-page book offers information on 2,500 products including electric vehicles, super-efficient lighting, and other options.

The Solar Electric Catalogue. Write to: Solar Electric, 175 Cascade Ct., Rohnert Park, CA 94928, or call 707-586-1987. A useful catalogue for solar-minded consumers.

Index

Acid rain, 32, 34, 51, 74
Aerosols, 13, 21–22
Agriculture, 4, 5, 19
Air conditioners, 13, 38, 52, 68, 82–83
 automotive, 68, 77–78
Appliances, 52, 54–55, 73
 energy-efficient, 86–87
 standards for, 55–56
Atmosphere, human activities altering, 1, 8–9, 14, 15
Automobiles, 15, 41–44, 46–47, 73, 76–78

Biological diversity, 5, 20, 34, 64, 65
Biomass, 13, 14, 58, 62, 69
Building codes, 56
Buildings, 52–57, 59, 94–95

CAFE (Corporate Average Fuel Economy), 42, 43

Carbon, 33, 34, 35
Carbon budget, 11–12
Carbon dioxide (CO_2), 6, 8–12, 16, 18, 25, 32, 91
 atmospheric concentration, 10f, 65
 global emissions, 11f
 individual responsibility for, 74–76
 taken up by oceans, 23–24
Carbon dioxide credits (proposed), 44
Carbon dioxide emissions, 1, 38, 42, 51, 71
 from buildings, 51
 reducing, 32, 33, 35, 36–37, 39–66
 from transportation, 76
Carbon-dioxide reducing technologies, 38
Carbon dioxide sinks, 24, 34, 64, 65

Carbon monoxide, 13
Carbon tax (proposed), 44
Chlorofluorocarbons, (CFCs), 8, 13–14, 25, 26, 32, 77, 83
 capturing and recycling, 68
 international agreement on control of, 27–28, 71
 reducing, 33, 34, 39, 66–68, 72
 substitutes for, 67
Civic participation, 95–96
Clean Air Act, 34–35
Clear air standards, 33
Climate change, 3, 5, 6, 12, 17, 18–20, 31, 66
 uncertainties about, 25–27
Clothes dryers, 88
Clouds, 22–23, 26
Coal, 6, 11, 27, 32, 35, 69, 74
Coal mining, 51, 69, 70
Computer models, 15–18, 22, 23, 24, 25
 global temperature increases, 27, 28f
Cooling, 52, 53, 73, 79–89
 solar, 58–59

Deforestation, 9–11, 24, 33–34, 35, 65
Developed countries, 38, 40
Developing countries, 2, 40, 63, 64
 technology transfers to, 71–72
Diet modification, 93
Dishwashers, 88–89

Economy (the), 32, 35–36, 40
Electric heat, 80
Electric vehicles, 49–50
Electricity, 5, 52, 62, 74
 consumed by lighting, 89–91
 water heater, 85
Energy audit, 84, 95
Energy conservation, 32, 35, 50, 53, 74
 government policy in, 56
 at home, 78–93
 tree planting in, 94
 at work, 94–95
Energy consumption, 40
Energy efficiency, 32, 33, 58, 64, 74
 in appliances, 86–88
 in buildings, 52–57
 and economic growth, 40
 in industry, 51–52
 least-cost planning in, 50–51
 technologies for, 39
Energy efficiency standards, 55–56
Energy intensity, 40
Energy production, 40–41
Energy sources, efficient, 2
 See also Renewable energy sources
Energy system, global, 28–29
Energy use, 39, 73
 in home, 78, 79f
 in manufacturing, 51
 in transportation, 41–44
Enhanced greenhouse effect, 1, 4, 24
Environment, 2, 32
Ethanol, 48, 58

Fans, 82–83
Federal Building Energy Performance Standards, 56
Feedbacks, 17, 22
Fertilizers, 14, 48, 70
Fluorocarbons, 67
Food Security Act, 64

Forests, 19, 72
    See also Deforestation
"Fossil air," 9
Fossil fuels, 9–11, 11f, 23, 32–33, 37, 42, 44
    alternatives to, 27, 58–62
    extraction of, 34
    taxes on, 44
Freezers, 68, 87, 88
Fuel economy standards, 41–44, 45
Fuel efficiency, 78

"Gas Mileage Guide," 78
Gas range, 89
Gases, atmospheric, 6
    See also Greenhouse gases
Gasohol, 48
Gasoline
    alternatives to, 47–50
General Circulation Models (GCMs), 15–18
Global warming, 1–2, 16–18, 20–21, 26–27
    basics of, 3–29
    contributions to, by country, 41f
    "High-leverage options," 32–36
    slowing, 29, 31–72
    uncertainties in, 21–27
Government policy, 34, 37, 38–39, 45, 56, 62, 67
"Green" technology, 72
Greenhouse effect, 1, 4–5, 6–15, 24, 63
    individual action in reduction of, 2, 73–96
    U.S. policy measures, 34–35
Greenhouse gas–emissions trading system (proposed), 72

Greenhouse gases, 2, 6–15, 19, 31
    global warming due to, 20, 26–27
    costs of controlling, 32
    global consequences of, 15–18
    need for international agreement on, 28–29
    reducing emissions of, 26, 35–36, 39, 66–70, 71, 93
    sources of, 6, 7t
Greenhouse problem, 12, 33, 67
    failure to act against, 31–32
    solutions to, 27–29, 32
    uncertainties in 24–27
Greenhouse science, 21–22, 24–27

Heating, 52, 53, 73
    energy conservation, 79–89
    solar, 58–59
Hot water, 84–86
Hydrocarbons, 15, 50
Hydrochlorofluorocarbons, 67
Hydrogen-powered vehicles, 50
Hydropower, 58
Hydroxyl radicals, 13

Individual action, 2, 39, 73–96
Industry, 51–52
Insulation, 37, 52–53, 83
Intergovernmental Panel on Climate Change, 16, 39, 71
International agreements, 27–28, 66, 71
International cooperation, 36–38, 70–72

Japan, 33, 37–38, 40, 43

Landfills, 13, 69–70, 92
Least-cost planning, 50–51
Light bulbs (lamps)
  fluorescent, 50, 53–54, 90–91, 95
  incandescent, 53, 89–90, 95
Lighting, 52, 53–54, 55, 56, 73, 95
  calculating cost of, 89–90
  energy conservation, 89–92
Local governments, 46–47, 70

Marine plankton, 23–24
Market mechanisms, 39, 44–45
Mass transit, 37, 44, 45, 46–47, 76
Methane, 8, 12–13, 26, 33, 37, 58
  control of, 68–70
Methane clathrates, 13
Methane emissions, 39, 48, 93
Montreal Protocol on Substances That Deplete the Ozone Layer, 28, 66, 72

National Appliance Energy Conservation Act, 55
Natural gas, 11, 13, 69, 85
  compressed, 47–48
Nitrogen oxides, 15, 32, 50
Nitrous oxides, 8, 14–15, 37, 48
  controlling emissions of, 39, 70, 93
Nuclear power, 37, 62–63

Oceans, 18, 21
  as sink, 6, 10–11, 23–24, 26
Oil, 32, 33, 51, 69, 70, 74
Ozone, 15, 21–22, 32
Ozone depletion, 14, 15, 74

international agreement curbing, 27–28
  reduction of, 66, 67
Ozone hole, 8, 14, 26, 27
Ozone layer, 13–14, 33, 66

Personal choices, 93
Petroleum extraction, 13
Petroleum industry, 51
  See also Oil
Photovoltaics, 58, 59–60
Policy scenarios, 27, 28*f*
Political action, 95–96
Pollutants, 2, 5, 50
Precipitation, 20–21, 25

Radiation, 8, 12–13, 33
Radon, 84
Recycling, 37, 56–57, 92–93
Reforestation, 64–66
Refrigerants, 13, 67, 68, 77, 78
Refrigerators, 54–55, 56, 68, 72, 86–88
Renewable energy sources, 32, 44, 50, 58–62, 64
Research and development, 38, 59–60, 62
Rice production, 13, 68, 69
Ruminant animals, 13, 68, 69

Salinity Gradient Solar Ponds (SGSP), 60
Sea level(s), 19, 21, 31–32
Showerheads, low-flow, 85–86
Sick building syndrome, 84
Sinks, 6, 8, 12, 13
  carbon dioxide, 24, 34, 64, 65
  oceans as, 6, 10–11, 23–24, 26
Smog, 5, 15, 32, 49, 51
Soil erosion, 65

Solar energy (power), 8, 23, 49–50, 58–60, 62
Solar output variability, 20
Solar ponds, 59, 60
Solar thermal systems, 58–59
Solid waste management, 57, 74
  See also Landfills
State governments, 47–48, 57, 70
Superinsulation, 52–53, 83–84

Taxation, 39, 43–45, 47
Technologies, 38, 39, 67, 71–72
Temperature increases, 3–5, 6, 12, 20–21, 25
  computer models, 27, 28f
  projected, 16–18
Thermostat settings, 80
Third World. *See* Developing countries
Trace gases, 6
Transportation, 41–44, 76–78
Transportation funding policies, 39, 45
Tree planting, 23, 34, 35, 65–66, 93–94

Tungsten-halogen lamp, 91–92

United Nations Environment Program, 16, 17
United States, 2, 33, 34, 95–96
  biomass potential, 58
  energy consumption, 40
  goals for, 38–39
  industrial energy use, 51
  leadership role, 31, 64–65, 70–71
  R & D spending, 62
  wind power, 60–62
Utilities, 50–51

Vegetation, 6, 64

Washing machines, 86
Water conservation, 86
Water heaters, 55, 85
Water vapor, 8, 21–22
Weatherizing, 79–80
Wind power, 58, 60–62
Windows, 54, 79, 80, 94
  modernizing, 80–82
Wood burning, 6